Knowledge Production Modes between Science and Applications 2

Series Editor
Jean-Charles Pomerol

Knowledge Production Modes between Science and Applications 2

Applications

Jean-Claude André

iSTE

WILEY

First published 2024 in Great Britain and the United States by ISTE Ltd and John Wiley & Sons, Inc.

ISTE Ltd
27-37 St George's Road
London SW19 4EU
UK

www.iste.co.uk

John Wiley & Sons, Inc.
111 River Street
Hoboken, NJ 07030
USA

www.wiley.com

Library of Congress Control Number: 2023949774

British Library Cataloguing-in-Publication Data
A CIP record for this book is available from the British Library
ISBN 978-1-78630-935-8

Contents

Foreword: Additive Manufacturing: From 3D Printing to Bio-printing vii

Introduction to Volume 2 xi

Chapter 1. Socially Responsible Research (SRR) 1

1.1. Introduction. 2
1.2. Setting the scene 5
1.2.1. Decision-making and ethics. 11
1.2.2. Technological progress: the researcher and the risks 19
1.2.3. Scapegoating 29
1.3. Modes of action 39
1.3.1. The project framework: socially responsible research 42
1.3.2. Other considerations 44
1.3.3. Socially responsible research 46
1.3.4. A few examples 53
1.3.5. Authoritarian constraint 60
1.4. Provisional conclusion 65
1.5. Conclusion: from the 12 “Labors of Hercules” to the 12 “death valleys” 67
1.6. References 72

Chapter 2. 3D, 4D and Bio-printing Innovations and Additive Manufacturing 91

2.1. Introduction. 92
2.2. Additive manufacturing or 3D printing 94

2.2.1. Invention of the additive manufacturing concept 96
2.2.2. "In-house" activities 103
2.3. Reaching out to society 106
2.4. Consequences 111
2.5. 4D printing 113
2.5.1. Reminders about 4D printing 117
2.5.2. Gaps between paradigms 125
2.5.3. Conclusion 127
2.6. Bio-printing 129
2.6.1. Principle of bio-printing 130
2.7. Discussion 138
2.8. Conclusion 142
2.9. References 147

Chapter 3. Creativity and Additive Manufacturing 161

3.1. Toward a Big Bang of creativity 162
3.1.1. Background and theoretical considerations 163
3.1.2. Introduction 164
3.1.3. Just before the idea 166
3.1.4. Clarification 170
3.1.5. The question of finalized knowledge integration 175
3.1.6. What interdisciplinarity to call upon? 178
3.1.7. Grains of sand in the clarification process 179
3.1.8. Conclusion 183
3.2. A comparison with 3D, 4D and bio-printing 186
3.2.1. Additive manufacturing 187
3.2.2. 4D printing 191
3.2.3. Bio-printing 198
3.3. Conclusion 202
3.4. References 205

Conclusion 215

Index 227

Foreword

Additive Manufacturing: From 3D Printing to Bio-printing

Jean-Claude André's new book is invaluable because it is written by someone who has a wealth of experience and has studied and followed the path from the researcher's idea to industrial application.

For over 50 years, Jean-Claude André has successively or simultaneously been a researcher, deputy scientific director of the CNRS engineering sciences department, scientific director of a joint organization devoted to workplace safety and advisor to the Institut national des sciences et de l'ingénierie des systèmes (INSIS-CNRS). It is fortunate that he thought to put his experience in writing because, unfortunately, in France, the invaluable experience of people like Jean-Claude André is often ignored by "managers" with no culture of innovation and often even without the slightest knowledge in science and/or engineering. May they one day open this book, promote innovation and halt the deindustrialization of France.

Jean-Claude André begins by reminding us of the role of manufacturing in a country's economy, and its dependence on science and innovation. He points out – and no one should ignore this – that a job created in an industrial zone, as a result of political incentives and subsidies, and inaugurated with much fanfare, costs the community at least 10 times more than a job created in an innovative startup.

A large part of the book (in two volumes) is devoted to the transition from invention to innovation to industrial manufacturing. Chapters 1 and 2 of Volume 1 follow the long and winding road from one to the other, from supporting creativity to mobilizing producers, via the indispensable POC (proof of concept). Jean-Claude André describes in detail all of the pitfalls to be avoided in transforming an idea into a product. We must thank him for shedding light on the decision-making process, the immoderate and sometimes ignorant use of TRLs (Technology Readiness Levels) by the research administration in France and Europe. He also devotes several pages to industrial property and patents, indispensable for private research, yet disastrous as performance indicators for researchers since they measure above all an institution's ability to pay lawyers and annuities. A great deal of space is devoted to the startups that line the road from idea to unicorn, many of whose corpses lie by the side of the road. We look at the various ways in which we can help them avoid going off the rails.

In Volume 2, Chapter 1 focuses on the social responsibility of researchers, inventors and companies. Communication plays a key role as an interface between science and the public. The difficulties and weaknesses highlighted during the Covid-19 crisis are well known. The author does not shy away from the issue, stressing the extent to which communication errors, ill-considered announcements made by ignorant people, the procrastination of expert advice and false hopes fuel the resentment of the crowd, always quick to look for scapegoats.

Beyond individual responsibility, Jean-Claude André advocates socially responsible research (SRR). At the end of the first part of this two-volume work (Chapters 1 and 2 of Volume 1 and Chapter 1 of Volume 2), the author summarizes in a table (section 1.5, Volume 2) that all of those responsible for innovation policy should have in front of them lists of the 12 obstacles to overcome in order to move from the researcher's idea to innovation and economic success.

Chapter 2 of Volume 2 looks back at Jean-Claude André's experience as inventor, with Alain Le Méhauté, of 3D printing, now known as "additive manufacturing". Jean-Claude André describes the genesis and development of his adventure, which led from a laboratory experiment to additive manufacturing and its various technologies. An adventure with sales in the billions of euros per year, which is far from over. It is a fascinating story that sheds light on the blindness of many institutions and officials who prefer to do nothing, rather than take the slightest risk of supporting an innovation that could ultimately fail. It is as if we should only support startups that we are sure will become unicorns!

From 3D printing, Jean-Claude André moves on to 4D, describing the problems and associated technologies. He explains how this type of innovation must initially rely on a niche policy. After 4D printing, we move on to bio-printing and its challenges.

Chapter 3 of Volume 2 provides an overview of all of the processes that lead from creativity to industrialization. Asking the right questions, valuing interaction between researchers, fostering interdisciplinarity and integration are just some of the recommendations that emerge, drawn from the experience of starting up 3D printing, without forgetting the role of chance. The future of 4D and bio-printing is considered in the light of the applications that are already possible and described.

All in all, this book contains the invaluable lessons of real-life experience and encyclopedic knowledge, backed up by a truly impressive bibliography. Nowhere else will you find such an important body of knowledge on the links between research, invention and industrialization. In addition, Jean-Claude André's erudition is also reflected in an impressive number of quotations, sometimes surprising or amusing, but always highly illuminating, making this book an enthralling read, dense but exceptional in its truth of a lifetime's experience.

Jean-Charles POMEROL
President of the Agoranov incubator

Introduction to Volume 2

Research management, the operation of projects, overly production-based evaluations, weaknesses in interdisciplinary project management, and the position of academic research in relation to the business world are all issues raised in Volume 1 and are reexamined in this volume, simultaneously addressing the somewhat uncertain and difficult link between ideas and innovation, a key element in the development of promising emerging technologies. This work, in a way a prerequisite to any link with society, is undoubtedly useful with its bottom-up logic, with a modest number of actors to save time in producing a convincing proof of concept. But this is clearly not enough, and this aspect is discussed in this volume.

The previous volume showed that there can be no innovation without society, especially not against society, because invention, as well as innovation, are aimed at society. However, before the "happy" idea and its users are connected in real life, there are many obstacles to overcome, often far removed from science, yet some of them are part of it. To take one example, scientific research brings to the forefront complexity in invention. It cannot stop there, because the downstream, productive sector needs clear procedures for making objects, with simple and robust processes resulting from this intellectual "fermentation".

On the one hand, we need to make radical transitions between complexity and determinism in order to produce but, on the other hand, attractive inventions are needed to make industrial sense of this production for interested customers. This means that, as soon as possible, these in-depth interactions between a science open to society and industrial production open to science must be put in place as soon as possible, despite the various "Herculean tasks" presented in Volume 1. They should lead to a genuine pooling of interests, and perhaps one day to a real merging.

In the long run, however, the relevance of a project will be judged by the researcher and/or their research unit on the basis of upstream risk-taking: individual risk-taking from the outset, exploiting the imagination, and then more collective risk-taking in scientific and technological support groups. This means that the organization of the laboratory must include reflexive, supportive partners who anticipate change, and who are not only committed to "protective values" (satisfying passive precepts of biased for example) or the hierarchical logics of an old order and ordinary careers...

This evaluation cannot therefore be solely academic in the disciplinary sense! In this context, science (researchers and organizations) must acknowledge its societal responsibility, which must be aimed at deepening and broadening knowledge. Science is not a trinket to be displayed in a global frame of reference; it has a local, or even wider, social or societal utility. It therefore needs to think responsibly about its aims and consequences (and it would be nice if it could think about the resources it devotes to them!). It must include a reflection on what technological progress represents, which is not an end in itself, but a means to living (well) on our planet. By reflecting with other stakeholders on its aims, it should not claim to be neutral, simply by asserting its objectivity, but should be committed to supporting the sharing of knowledge and the applicative powers associated with these achievements. Generally speaking, this is what Volume 2 aims to show, with the aims of responsible research, and from a more factual standpoint, around known situations arising from three-dimensional printing.

1

Socially Responsible Research (SRR)

The implementation of scientific advances in our social practices without us being truly attentive to their impact on our values and representations, or at least with the feeling of powerlessness to control the inevitable process.

(Hirsch 2021)

I think we were so happy to develop all this criticism, because we were so sure of the authority of science. And that the authority of science would be shared because there was a common world. [...] We didn't even need to articulate this notion of a shared world, because it was self-evident. [...] Now we have people who no longer share the idea that there is a common world. And this changes everything, of course[1].

(Latour 2018)

The eternal question of disciplines and professions for which it is difficult to separate the context of discovery and the context of justification.

(Lénel 2014)

For a color version of all the figures in this chapter, see www.iste.co.uk/andre/knowledge2.zip.

1 Author's translation from the French version of the text.

1.1. Introduction

With the atomic bomb, after Oppenheimer, the academic world questioned the responsibility of researchers and engineers for technological innovations. It is always difficult to distinguish between right and wrong... but, for example, when we know that 24% of greenhouse gas emissions in the United States are directly linked to the industrial sector, it seems worthwhile to examine how science can help reduce this rate (Everett 2022). In elongating this need for responsibility, Grundwald (2014) points out that questions of technology assessment today extend as far as its future consequences and the need to take account of society's expectations. However, these expectations are subject to change, as can be seen today on certain issues associated with the war between Russia and Ukraine. It is becoming increasingly complex to distinguish between the individual and collective responsibilities of those involved in innovation. Debref et al. (2019) talk about inscribing social responsibility into the innovation process:

– taking into account potential impacts on social well-being (Pavie 2012);

– using social and environmental constraints to drive innovation (Temple et al. 2018). In other words, innovation has thus become responsible.

For Saint-Simon (1923), it is a question of "revising all ideas to base them on the principles of industry, to relate all morality to production", a morality centered on satisfying all the needs of the members of the social body. Today, disruptive innovations are taking hold, notably with digital technologies associated with the concept of Industry 4.0. They are changing concepts, knowledge, our way of living together and therefore our social practices as a whole. "Every technical object is a negotiation [...]; it is simultaneously the contract that momentarily seals the balance of power and the weapon brandished for other conflicts and negotiations" (Roqueplo 1983).

Even if the exercise is a delicate one, an ethics of the present day should explore new scientific fields and their possible consequences for society. It should go so far as to envisage modes of consultation with society, with a view to making the necessary trade-offs, albeit provisional and incomplete, choices (Hirsch 2021). Innovation is exploding in many fields to the point where attempts at exogenous anticipation, and even monitoring, are proving impossible. This is the current challenge facing ethical thinking. From now on, researchers and creative thinkers have a responsibility linked to their projects to produce new knowledge that can have an impact on society: pluralism of viewpoints and pluridisciplinarity of expertise with a view to the common good.

"Without the freedom to criticize technology, there is no 'technical progress' either, only and when this conditioning becomes cybernetic, as is the case today with new technologies, the threat becomes considerable" (Virilio 2001). Virilio takes up the Kantian theme of the public use of reason, which must always remain free (Kant and Mendelsohn 2007). But with big science, and its successive transformations, the novelties enabled by scientific developments bring us new technological possibilities every day. According to Fukuyama (2002), "today there are too many commercial interests and too much handling of money for self-regulation to continue to function properly in the future". Similarly, researchers are guided in their choices by the possibilities of access to funding, directed by the authorities in major programs (ANR in France and Horizon Europe in Brussels).

So we need to discover new ways of paying attention (getting out of our ivory towers), sharing knowledge (between scientists and technologists, on the one hand, and with society on the other) and interdisciplinary expertise in increasingly complex and wide-ranging fields, mobilizing and affirmation of choices. "It's all about recognizing the interconnectedness and importance of situated knowledge, and not pitting sources of knowledge against each other and the knowledge sources of scientists" (Lénel 2014). This principle of socially responsible research should enable us to move away from an overly disciplinary and "sacralized" conception of science, specific to France. In these circumstances, for informed debates, according to Jaeger and Mispelblom-Beyer (2014), the following two conditions are necessary:

– to not regard non-ideologized disagreements as a waste of time, an obstacle to the emergence of a truth (at least provisional) that can simply be revealed by asserting arguments of authority;

– to avoid bypassing them with a "conciliationism" that avoids taking a position and leads to inaction (Charlot 1995).

NOTE.– Ethics and responsibility.

In principle, "research ethics" concerns the conditions of acceptability of research involving humans, society and animals. In this chapter, proposals will focus on the social impacts of innovation and little on human-related aspects, and even less on animal ethics (see, for example, Potter 1971; Richmond 2000; Pinsart 2009; Halioua 2017). Under these conditions, thinking does not consist of denying the existence of reality, of "facts", but in emphasizing the part of construction in the perception we have of them. More than the question of discipline, it is the question of the need to study action that emerges, as this is not the prerogative of any discipline, human sciences or so-called "hard" sciences. So, rather than talking about ethics, which could lead to confusion, we have chosen to talk about the principles of responsibility.

NOTE.– Acceptability.

The expression of a goal to better respond to user aspirations from the earliest stages of innovation, and not in the final phase, when work is limited to the social acceptability of the final product, corresponds, according to Chouteau et al. (2020), to the final exploration of the theme. But in both of these “considerations”, the word acceptability, more or less social, has been used. By “contribting to a metaorphosis” in the way some people deal with this word in relation to society, the discourse on responsible innovation has generally avoided undergoing a thorough conceptual deconstruction (Thapa et al. 2019). However, the dictionary indicates that acceptability refers to a character of something that is more or less tolerable. It is a tricky concept to define, apart from the perception, founded or unfounded, of risks. Since the end of the 30 years post World War Two, in most Western countries, public protests have been voiced. The rise of these reactions associated with technologies appears to be a public thematization of risks, that is, the formation in the public sphere (in the Habermasian sense) of universalizing norms such as “environmental and health protection” (Lehoux et al. 2019).

Generally speaking, the French are clearly interested in new technologies (Bauer et al. 2021). But within this framework between the social body and the worlds of research and innovation, how do the players – all of the players – position themselves, and who is ultimately responsible for what?

The necessary relationship between science and society undoubtedly need to be reexplored periodically with a view to strengthening trust between all stakeholders. The malaise linked to the loss of trust in the “system” is reflected in a number of opinions; but is it possible to find the right words to express and operationalize this search for agreement between science, technology and society (Kerninon 2009)? Ronsavallon (2006) distinguishes several forms of democracy:

– involvement where citizens join forces to produce a “common world”;

– expression, linked to the manifestation of collective feelings;

– intervention that aims for joint action to achieve a chosen result.

In these three forms of political activity, it is clear that the “involvement” component of innovation is not strictly covered (Lechevalier 2019). Forms of mistrust and loss of legitimacy of certain technologies emerge from this deficit. There are several reasons for this negative assessment:

– the lack of explanation of what is being promoted in technological terms;

– the absence of an overarching vision;

– the weakness of a common intellectual elaboration between public and experts;

– the development of a "social disarticulation" by not sufficiently associating technological advantages and risks for humans and the environment (Touraine 1988);

– abusive extension given to the notion of harm (degradation of the principle of "non-negligence" according to Ogien (2007)).

1.2. Setting the scene

> Our culture dictates that we must win: on the battlefield as well as in cut-throat economic and scientific competition... now that we've reached the extreme limits of performance, martial violence and economics, are we so sure that we must always win, even spiritually? (Serres 2007).

> Are innovations revolutionary? Are they based on originality, the surprising, the unexpected? On the contrary, most of them are very conventional. They're based on simple, predictable laws of evolution that nature itself knows how to put to good use. To innovate is to think outside the box (Dortier 2015).

> The greater the speed of operation, the slower the response to the unexpected event (Ellul 1988).

The company creates technology in relation to its own culture and its own traditions, and in a recursive manner, technology is an important element in social evolution, "as customs adapt to technical availability and constraints and this transformation is often unexpected" (Gaudin 2005). Thus, it is becoming increasingly clear that man-made objects are not neutral and can lead to forms of ethnocide through artifacts (Jaulin 1970, 1977) but science, technology and society must integrate human components into all of their psychological and social complexities, as well as their temporal evolutions (Barré 2020).

On this theme, Habermas (2002) writes:

> At first glance, moral theory and ethics seem to be preoccupied with the same question: 'What must I do?'; 'What must we do?'. Only 'ought' no longer has the same meaning when we ask ourselves about the rights and duties we impute to everyone reciprocally from the point of view of the first person plural, and when, concerned about our own lives, we adopt the point of view of the first person singular to ask ourselves what, all things considered, is better to do in the long run 'for me' or 'for us'.

Thus, the purpose of this chapter is to introduce the framework of responsibility that could/should inhabit the thinking of researchers and/or inventors, especially if we take Michel Serres' (2008) opinion on the relationship between technical innovation and society into account. He writes: "By separating nature and culture in this way, have we committed an error of judgment, resulting in a mortal crime against ourselves and the world, both inert and living?". Indeed, in attempting to answer this question, we must take into account the influence of the progress of scientific knowledge on the understanding that researchers/inventors have of themselves as people responsible for their actions, even future ones. This reflection on responsibility corresponds to a modest proposal for a commitment to scientific research, adapted to "techno-sapiens" (Lipovetsky and Serroy 2008), toward an "ethics of knowledge".

As Figure 1.1 shows, the TRL values of these ideas, if possible prior to the production of a POC, are quite low.

Although the focus is on making the outcomes of innovation processes more accountable and desirable, the technological and commercial natures of these processes are not questioned in this book (for more on this topic, see, for example, Berger-Douce 2015; von Schomberg and Blok 2021).

Responsible innovation refers, according to Lehoux et al. (2019), to "different types of socio-technical systems that can be described as 'responsible' because they share the objective of meeting societal challenges and are oriented towards a common good that recognizes that life in society depends on the good governance of a shared set of resources (material or not)" (Flahault 2013; Forsyth and Johnson 2014). It involves an ethical reflection on the merits of technological development in a given social framework (von Schomberg 2013; Stilgoe et al. 2013).

Figure 1.2 illustrates the main components of the Lehoux et al. (2019) model and summarizes the empirical categories derived from individual studies. In this model, the rather interdisciplinary cognitive-technological activity involves different categories of resources of various origins (materials, capital, technologies, practical or scientific knowledge, etc.). This activity proceeds by "creative DIY" (Gurca and Ravishankar 2016), mutual learning and experimentation (see POC). It requires a technological and ethical design flexibility to envisage a socially "acceptable" achievement. Even if this is not part of French culture, it could be fostered by a cultural context where failure is seen as a source of learning and where the innovator may have room to maneuver. This is all the more true in the early stages of POC (Wittrock et al. 2021).

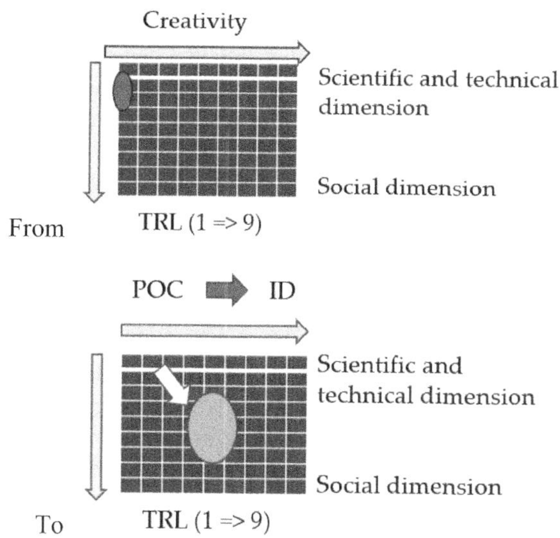

Figure 1.1. *TRL spaces for SRR*

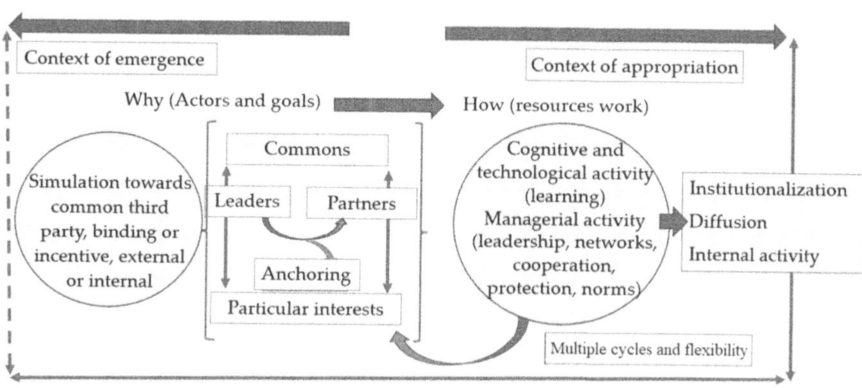

Figure 1.2. *Socially responsible research (SRR) between creation and society*

Without it being easy to notice, artifacts act through feedback on practices and culture of society to such an extent that de la Miranda (2010) has written: "By acting on matter, Man simultaneously acts on himself. This makes us not just external users, but theoretically responsible creators, insofar as technology is a part of ourselves". Within this framework, Attali (2007) assigns a number of missions to the world in the making. He writes:

> Freedom in the sense of less pain and effort, more free time, despite the eminently illusory nature of this freedom 'enclosed within four walls that are those of death and birth'. This freedom, the desire to live better within these four walls, the will to remove these walls by making life longer and better, and then the will to pass on better living conditions to those who will later live in this same room.

This ethic on how to "live well together", in the sense of Ricœur (1990) ("aiming for a good life, with and for others in just institutions"), leads sciences and/or creative people to help produce innumerable goods, elements of technological progress. In fact, today it seems that our technological know-how is increasingly outstripping man's knowledge, which is being disrupted by the very rapid dynamics of change. In any case, Puech (2008) writes that "technology is only neutral if it is considered as a means only, the difficulty being in the 'only'". Indeed, it cannot be considered neutral because of the instigation associated with the production of new consumer goods, and, as Ellul (1954) points out, because of the "principle of irreversibility of efficiency".

Feenberg (2004), drawing on cultural and ideological presuppositions, proposes that the approach to science-society implications, in relation to the technological adventure and its social significance, can be summed up in the following propositions:

– The technical design is not determined by a general criterion of efficiency, but by a social process that selects between technical alternatives according to a variety of case-specific criteria.

– The social process does not concern the satisfaction of "natural" human needs, but the cultural definition of these needs and therefore of the problems posed to technology.

– Competing definitions reflect conflicting social visions of modern society, embodied in different technical choices.

But what becomes of these proposals if we consider Saint-Exupéry's views on a form of soft totalitarianism? In 1943, he wrote of France and the United States:

> Today's man is held together, depending on the environment, with belote or bridge. We are surprisingly well castrated. So we are free at last. Our arms and legs were cut off, then we were left free to walk. Personally, I hate these times, when mankind is becoming gentle, polite and quiet cattle under 'universal totalitarianism'. We're led to believe that this is moral progress [...]. People who are fed ready-made food, standard food, like cattle being fed hay. That's the man of today (Saint-Exupéry 1956).

This aspect of social control is an essential part of SRR thinking and will be examined further below.

Heidegger (1958) wrote: “What kind of decision-making mechanism be put in place to enable all citizens to express their views, not on research itself, but on the questions raised by the applications of research?”. The rapprochement and integration between science and innovation leads us to reflect on the responsibility of the researcher or creative artist, in a context marked by an increasingly ambivalent acceptance of risk. While the researcher and the “applicator” are bound by forms of responsibility yet to be defined, the user is not always innocent of the use he makes of the devices he acquires. This is the case with the numerous detours of innovations that we have seen that had totally escaped the designers... In case of the exploration of uncontrolled fields in terms of risks to society and the environment associated with technological research, we need to explore new forms of social and cultural rationalization.

So, according to Neubauer and Joly (2004), there is a form of social control, the need for direct in scientific choices (see also Dietz 1995; Irwin 1995; Dienel and Renn 1997; Andersen and Jaeger 1999; Calame 1999; House of Lords 2000; Callon et al. 2001). The advantages of these practices can be summarized as follows:

– Participatory procedures can contribute to the political and objective legitimacy of decision-making processes. If we follow Callon’s (2003) definition of objectivity: “Knowledge can be said to be all the more objective when it has been put in a position to respond to the greatest possible number of objections”, then decisions taken with the participation of many different players have a better chance of being accepted by the public. The problem is to know how to share honestly between the knowledgeable and the uninitiated, and how to make the issues understandable to the public. Participation should, however, create a proximity between politicians and citizens that should reinforce the transparency of decision-making processes and encourage the reinstallation of trust.

– Citizen participation should improve the knowledge base on which decisions are made, as it includes the diversity of knowledge (local, empirical, traditional, and different worldviews) that is taken into account alongside expert knowledge (and which forces the “translation” of confined knowledge), enabling solutions to be found that are better adapted to the complex realities on the ground (Callon 2003; Neubauer and Joly 2004).

– The open science movement (Nosek et al. 2015) represents an effort to resolve the credibility crisis of coupling by promoting transparency in the research process (Tsui 2021).

– "Citizens' conferences or juries are places where new forms of knowledge creation are practiced. Unlike a poll or plebiscite, questions are open-ended and new ideas can enter the process".

– The use of expertise, in a very broad sense and across very different professions and sectors, ensures a multi-disciplinarity of viewpoints, values and different elements (scientific advice in politics was based on the idea of the independence of experts). However, this independence has become problematic, not least because of the commercialization of research.

– Deliberative procedures must avoid the predominance in decision-making of the particular of a single group of players. Experience proves the competence of laypersons in dealing with complex issues, and their ability to act in the general interest and from a long-term perspective.

– One of the aims of participatory procedures is also to enlighten citizens (without actually sharing the "deficit model") and thus contribute to opinion-forming. However, this transfer of scientific knowledge should take into account the variability of "lay" audiences who need to form their own viewpoints (Allum et al. 2008; Schefeule et al. 2009; Jakobsen et al. 2019; Wittrock et al. 2021; Deserti et al. 2022). What is more, this plurality of viewpoints is hardly diminished by the media industries that would, according to Brumfield (2009), be at the origin of reliable sources of information adapted to citizens' needs.

– From an economic point of view, although sometimes expensive at the outset (carrying out a procedure), they have a very good benefit/cost quotient over the medium and long term, since they have the advantage of preventing potential conflicts, anticipating resistance upstream and making the necessary adjustments in good time.

– From a practical point of view, co-created knowledge is needed to envision multi-dimensional and multi-scale innovation dynamics, initiated and monitored to encourage the creation of economic value alongside solutions to different challenges. In these processes, it is necessary to "test the innovation" to guarantee accountability of the final product.

– Scientific skills and competencies required for innovation as an exploratory mode covering the central theoretical perspectives and fundamental concepts of responsible innovation, on the one hand, and empirical insights into how actors address dilemmas and related tensions in different contexts when innovating, on the other hand (Jakobsen et al. 2019).

Schwarz and Thompson (1990) define technological development as a dynamic, evolutionary process that associates technologies with patterns of organization, even disorganization, in conjunction with creative ideas, thus introducing a three-dimensional perspective to studies of responsible innovation (see also Fløys and Jakobsen 2017; Sjøtun 2019). Figure 1.3 therefore takes into account the fact that emerging innovations are part of a more complex discursive process, insofar as they are likely to produce a dynamic shift in "rationalities".

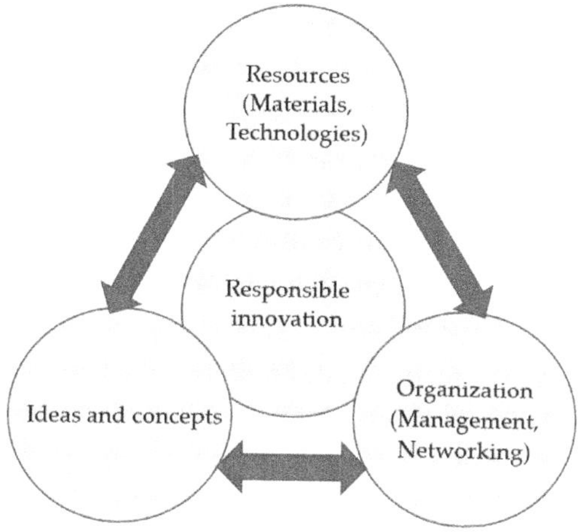

Figure 1.3. *For a responsible innovation*

1.2.1. *Decision-making and ethics*

> Compartmentalizing disciplines leads either to literary people who don't know anything about science, who are ignorantly cultured, or to scientists who don't know anything about literature, who are uneducated and uncultured, i.e. two varieties of imbeciles. The problems we face today require all disciplines (Serres 2015).

> The rejection by scientific thought of all ideas based upon value concepts, such as perfection, harmony, meaning and aim, and finally the complete devalorization of being, the separation of the world of value and the world of facts (Koyré 1968).

> A society's organizing system co-evolves with its production system and defines how it understands the world and governs behavior, encompassing models of thought, belief systems, social systems, political systems, economic systems, and governance structures which impact ways of thinking, seeing and making decisions at individual, institutional and collective scales. The ultimate design of the production system will be indelibly based on choices and decisions

relating to how we design the different elements of our organization system (Arbib et al. 2021).

Throughout this chapter, we will examine the difficulty of putting into operation the various aspects associated with the "socialization" of scientific activities, the researcher's "freedom", which makes them responsible in a democratic society because, for various reasons, they are obliged to make choices. In our Western democracies, individual freedom is less a right than a duty. The fact that it can be considered illusory in the name of all the determinations over which we have no control, and which guide our decisions, does nothing to change the imperative it indicates (Jurdant 1994). In another context, Comte-Sponville (2004) defines, in relation to decision-making processes, four interacting orders of potential interest to the world of technoscientific research:

– The techno-scientific order: we need to define the boundary between what is scientifically possible and what is impossible (this means that if only this order existed, as long as all of the funding was available, everything possible would normally be explored).

– The legal-political order corresponds to the democratic order and may differ from one country to another. The state manages constraints and interests to ensure the stability of the social body. In principle, there is no moral position on the part of the state in democratic systems.

– The moral order: this order subjects researchers to a certain number of moral requirements (if only because the knowledge of scientists, in the fields they dominate, is more important than that of citizens).

– The ethics order: framing actions with a view to the overall well-being of citizens (right/wrong relationships); this relationship to ethics depends on the society in which we live (evolving nature of perceptions, acceptability, attractiveness, context, etc.). It is an approach that aims to understand ourselves to know what is desirable or undesirable, good or bad.

The last two criteria focus on analyzing the moral justification for research into a new technology in processes of critical foresight, because the technologies in question are rarely value and impact free. In fact, being responsible means being able to make a decision in situations of uncertainty and, in particular, when it involves several orders at once, without confusing them. When they are in contradiction, responsibility is linked to the decision to give priority to one or the other. According to Simon (1947), the decision-making process is based on the representation of the "dissonant" situation in which we intend to intervene. It is conceivable, in line with Boudon (1977, 2009), that these more or less collective decisions will bring together different, even contradictory interests, but adapted to the situation (between a choice using "individual" decision theories or those of games taking into account the effects of interdependencies and the "unpredictable" behavior of others).

This gives rise to forms of fragility that need to be addressed as scientific activity progresses. And this is precisely what the notion of socially responsible research expresses (André 2013). The European Union has introduced a number of "constraints", for example in the European Charter for Researchers (Horizon 2020 2020), which brings together elements of deontology and ethics research. In the 19th century, De Tocqueville (1981) wrote:

> Not far from this class is another party, whose object is to materialize mankind, to hit upon what is expedient without heeding what is just, to acquire knowledge without faith, and prosperity apart from virtue; assuming the title of the champions of modern civilization, and placing themselves in a station which they usurp with insolence, and from which they are driven by their own unworthiness.

Figure 1.4, taken from Tijdink et al. (2021), gives an idea of the scope covered by the SRR concept.

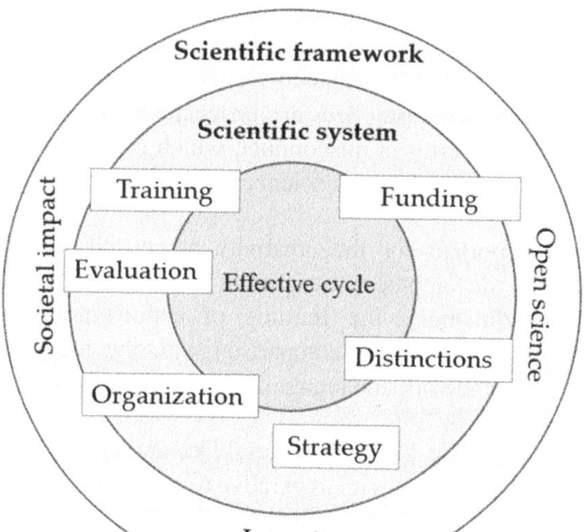

Figure 1.4. *Schematic representation of the elements that interact in responsible research practices; the three rings represent the different levels of consideration of SRR: scientific frameworks, the scientific system and the empirical cycle*

These authors have identified several main themes that need to be taken into account in future research in the field of social responsibility:

– evaluation responsible for research and researchers;

– the effect of open science and transparency on SRR;

– research on mentoring, supervision or responsible management and modeling managers;

– the effect of education and training on SRR;

– verification of reproducibility or robustness of results;

– mastering integrity challenges (Haven et al. 2019a, 2019b, 2021; Horbach and Halffman 2019);

– responsible evaluation.

NOTE.– Integrity.

Research practices that are, to say the least, imprecise from an ethical point of view, the research system with its high level of competition, unidimensional evaluation criteria, an individualistic research culture individualistic research culture and the pressure to publish for an academic career, can lead to unnecessary publications (Fanelli et al. 2015). Behaviors such as falsification, plagiarism and questionable research-checking practices are unacceptable (John et al. 2012; Haven et al. 2020). Research can result in misconduct, which can lead to rejection (Davies 2019) and risks of loss of confidence in science.

The situation is important for the creativity associated with artifact sciences, because "technological development transforms the meaning of what is human" (Heidegger 2003). Surprisingly, the framing of actions aimed at technological development is poorly developed in comparison with the life sciences, which are subject to a clear legal framework and recognized bioethical aspects (UNESCO 2005).

By way of example, Article 4 of the Universal Declaration on Bioethics and Human Rights (voted unanimously) imposes an imperative respect which illustrates this desire:

> In the application and advancement of scientific knowledge, medical practice and associated technologies, the direct and indirect benefits to patients, research participants and other affected individuals should be maximized, and any harmful effects likely to affect such individuals should be minimized.

Freud, in 1908, introduced elements now known as (Freud's) three frustrations: the first related to Galileo moving the center of the world; the second to Darwin placing the ape as Man's cousin; and finally, to Freud himself placing human behavior in a new context. Should we add to these losses of egocentricity a fourth frustration for the researcher involved in the technological future, linked to a questioning of the premise of science? The questioning of his "freedom" or power?

The aim of this chapter is to show that the main role of science, which was to draw a coherent and acceptable picture of the world in which we are forced to live, is changing. Since, in particular, Isaac Newton, the world of intellectuals has been dominated by "natural laws", based on rational, scientific, mathematizable and, until recently, essentially deterministic thinking (André 2008). On these foundations, Weber (1985) considers that, in modernity as currently constructed, the rational, dominating context of technology translates into a control of social life that threatens non-technological values. This tendency leads to what he calls the "iron cage" of bureaucracy. He is joined on this point by Taylor (2001), who characterizes our techno-scientific civilization by the arrival of "moments of complexity" which are likely to lead, without us being careful, to very profound transformations in society (which he defines as a transformation of our cultural plausibilities). Similarly, Janicaud (2002) further underlines this when he writes: "The danger in talking about techno-scientific societies is to suggest that we are sacrificing ourselves to a kind of 'ontological majoration' of tech-noscience, i.e. that we are betting on its capacity to found a culture, civilization and humanity". Puech (2008), for his part, believes that "we are absorbed by the objects and their petty concerns. For this reason, we are deprived of the dignity of a way of thinking that asks questions about Being, which nonetheless constitutes a characteristic potential of the human being". This futility of things constitutes, according to this author, what Heidegger calls "dereliction" (see Heidegger 1927; quoted in Puech (2008)).

But it was the success of the scientific method that brought society to this high technological level:

> If the human condition consists in Man's being a conditioned being for whom everything, given or man-made, immediately becomes a condition of his further existence, then man 'adjusted' himself to an environment of machines the moment he designed them. [...] Even the most refined tool remains a servant, unable to guide or to replace the hand. Even the most primitive machine guides the body's labor and eventually replaces it altogether (Arendt 1968).

In fact, the line between acceptance and rejection of a technology depends on whether the technology is considered neutral, value-laden or simply symbolic:

> The invention of the steam engine and the locomotive is tantamount to the invention of a rail disaster [...] Each period of technical evolution brings with it its own set of instruments and machines, and the

> appearance of specific accidents, revealing 'in the negative' the rise of scientific thought (Virilio 1996).

There are two at least partly separate areas to explore: ethics and/or morals, on the one hand, and risks, on the other hand.

So, crisis after crisis, emotion after emotion, we must now ask ourselves what place the "arrogant intelligence" of scientific rationality should have when, for example, reserves are dwindling or when the social order is profoundly challenged by its technological contributions. Technological innovation results in new, previously unimagined possibilities, and makes certain tasks easier to perform (Reschler 1998). However, if we think about it, the overall effect is to make people's lives more subject to progress and more complicated. How far can we go to avoid going beyond the simple conservative tensions associated with technological progress? According to Nordon (2006), some scientists, particularly in theoretical disciplines, "consider that knowledge does not have to be useful, and work 'for the honor of the human spirit'". The thesis of the practical neutrality of science, reduced to mere knowledge, avoids the realm of action. Others, emerging from a set of fallacious representations of purely contemplative scientific knowledge (Lavelle 2007), want to be useful and proclaim that fundamental research is indispensable to applicable research (and vice versa)? For the latter, intention, decision, actions and their consequences fit perfectly into responsible action.

"Philosophers distinguish between knowledge and action, usually on the basis of Kant's critical division between the faculty of knowing and the faculty of desiring" (Lavelle 2007). And what many of them have in common is a desire to have it both ways: to work solely according to their own personal tastes, without taking orders from anyone else, and to derive additional glory from the fact that their work sometimes has positive applications. As Stengers and Bensaude-Vincent (2003) remind us:

> The contradiction between the admission of a strategic interest and the exaltation of disinterestedness is carefully avoided. Better still, any question raised about the legitimacy of such a disinterested pursuit is regarded as an act of criminalization – even a manifesto of anti-science.

This desire to be able to satisfy our own scientific desires is naturally also found in academic laboratories oriented toward engineering, as in all of the other sciences more specifically oriented toward the deepening of knowledge: "In many respects, society recognizes the importance of a search for knowledge [...]. However, it is natural that there are more specific expectations on certain subjects and with a view to certain objectives" (Prieu et al. (2004); see also Forman

(2007)). However, lost in his solitude and in the multitude of disciplines, the researcher, in good faith, may never realize the influence he exerts. In 1819, Benjamin Constant, on this subject, but in a different context, wrote the following eloquent sentences: "Never did his will imprint itself on the whole: nothing establishes his cooperation in his own eyes [...]. We have lost in imagination what we have gained in knowledge" (Constant 1997).

NOTE.– Interdisciplinarity, science and responsibility.

> How can we 'share' (in the etymological sense of the word communication) knowledge and build scientific knowledge off the beaten track? Our answer is twofold: by developing a dialogical ethic between indiscipline and responsibility, which is only possible by adopting an epistemology that breaks with the dream of a reason capable of saying what is true in an absolute and total way. In other words, drawing on the anarchist lessons of Kropotkin and Feyerabend, for whom it is the search for the researcher's emancipation that must govern the construction of scientific knowledge, and not confinement in some absolute and unsurpassable 'truth program' (Veyne 1983) – whether the latter is called 'discipline', 'pluri-discipline', 'inter-discipline' or even 'indiscipline'! (Dacheux 2013).

Classical science cannot think about its social and ecological responsibility because "for there to be responsibility, there must be a conscious subject; but the classical scientific vision (deterministic and reductionist) eliminates consciousness, eliminates the subject, eliminates freedom [...]". This "ethical blindness" of science is also due, continues the sociologist, to "the disciplinary culture, which fragments knowledge, and specialized training, which renders the scientist ignorant of, and then indifferent to, epistemological and, of course, ethical issues. Science is blind to what it is, what it is becoming, what it could or should be" (Morin 2004).

But under these conditions, society is increasingly calling to account those who have led it to its current state, unquestioningly dragged along by the media and advertising to the inevitable technical progress (to make themselves understood or to mask problems, such is the new "law" of communication (Mercier 2008))!

According to Tubiana (2007), "even the most rigorous scientists are afraid of upsetting the media by daring to tell the truth, because they feel that one day they will need them, so it is better keep quiet rather than oppose 'political correctness'". This situation of formal imbalance allows active groups (elements of temptation of democratic populism?) to unrelentingly raise the question of whether technology has indeed been developed to the measure of Man, in order to serve some purpose, or to less obvious objectives... However, "the social distribution of knowledge of certain elements of daily reality can become highly complex and even confusing for the layman […]. In such a case, I seek not only the opinion of experts, but also and above all, the opinion of experts about experts" (Berger and Luckmann 2005). How, on such a basis, can trust reign? Thus, a loss of confidence develops in society, which, at the same time, demands novelty, which still allows us to affirm that if the scientific method and technology have conquered the world, it is because they tautologically correspond to positive and useful forces.

Technological evolution has a number of specific characteristics: it is irreversible, irresistible, discontinuous and uncontrolled. However, for a long time, the consequences of new technologies on society have only been considered as second-order consequences, since the effects are not related to the technology itself, and applications are at best the subject of general moral or ethical analyses. In reality, technological advances are essentially the object of at least apparently continuous processes, but they colonize our lives and insensibly change our habits and behavior (Ten Have 2007). This evolution does not normally give rise to ethical or moral questions, unless, through triggering mechanisms of various origins, they reveal the crossing of thresholds considered unacceptable. Minguet and Osty (2008) point out that:

> radical uncertainty refers to situations in which actors are unable to draw up reasoned anticipations of their actions, or to establish eventualities or probabilities concerning what might happen. They are faced with the impossibility of strategic calculation.

So this increasingly rapid transformation of the world translates a posteriori into tensions, constraints, unhappiness and profound cultural changes in society as a whole (Jacob 1997). This leads the social body, reinforced in its concerns by the power of the media, to look to science and technology for the causes of the malaise and to question those responsible to take a less specialized, more holistic stance (UNESCO 2000).

In this disintegrating landscape, is it possible, as Gonod (1990) puts it, "for society to master the development of the technology that is its creation and whose almost carcinogenic growth threatens its very existence, at the same time that it creates tremendous opportunities for humanity"? This chapter proposes the modest exploration of a "new" framework for scientific action which should make it possible, perhaps better, to redefine the role assigned to science and invention in our world in crisis. To achieve this minimum objective of pooling points of view, it is necessary to aim for bringing together science and technology, on the one hand, and social, cultural, political and economic realities, on the other hand, in order to achieve progress resulting from creativity and science, useful for society.

1.2.2. *Technological progress: the researcher and the risks*

> We are supposed to be trained for society and we are instructed as if each of us were to spend his life thinking alone in his cell, or dealing with indifferent people on trivial subjects (Rousseau 2009)[2].

> Interdisciplinary exchanges are not random. They can only be understood in terms of the subversive or legitimizing role they play in the host discipline. Thus, the chronology and modalities of transfer depend first and foremost on the conjuncture of the receiving milieu [...]. Like translation, the practice of interdisciplinarity is always a partial betrayal (Grenier and Lepetit 1989).

> Pluridisciplinarity, while it encourages confrontation and respects the pluralism of opinions and judgments, nevertheless remains within the strict disciplinary framework that was at the origin of the request. This is why the multidisciplinary approach, even if it provisionally replaces the compartmentalized disciplinary field with a communicational inter-field, runs the risk of resulting in a juxtaposition of points of view, with no overall coherence (Resweber 2000).

Prometheus, definitely unleashed, benefiting from the development of scientific knowledge and its achievements in terms of innovation, producing new things faster and faster, engages society as a whole in an indefinite process that leads to a plurality of unknown futures. These include those which presuppose the possibility of remaining within old, tried-and-tested paradigms, or, on the contrary, new areas for reflection and/or action based on foundations not yet taken into consideration (augmented man, new applications of genetics, sustainable development, global warming, reinforcement of the primacy of technology, intercultural and even inter-religious issues, etc.).

Having lived for millennia by following/subjecting time and the seasons – that's the whole agrarian period – mankind has almost succeeded in reversing the trend. Thanks to technology, it has even been possible to believe that neither time nor the seasons are any longer submissive, but increasingly controlled (alas, global warming is here to undermine this desire). Boissonnat (2007) reminds us that money is one of mankind's three great passions, along with sex and power. This change in temporal scale is associated with others: globalization, the apparent mastery of complexity through the fragmentation of production and, to a lesser extent, interdisciplinarity. Everything around us is on the move: the apparent, such as technology and its manifestations, but also the transition from work to employment, to precariousness

2 Author's translation from the French version of the text.

and the "obligation" to be free. The rules that lay bare the lives of so many people are a form of destructuring arbitrariness, usually far removed from the slightest democratic debate. As Durkheim (1963) already pointed out, "the republic is just when it ensures that contracts between individuals are truly free, that is, when it can certify that the contracting parties are placed in equivalent situations". However, the only thing that can be analyzed in this respect is the permanent movement that characterizes virtually all social structures. In this context, the problem of permanent movement has replaced that of simple change.

Technical innovations in all their forms are a silent revolution disrupting the state, the citizen, the employee, the hierarchical system within the company and the company itself, which is shifting from its role as vertical integrator to that of modular production, as a result of fragmentation... According to Janicaud (2002), technological progress is a dangerous utopia, because it provides ever-renewed material goods for the immediate satisfaction of "false" needs, whose long-term consequences are masked. To achieve this, companies are forced to develop their performance (from the individual to the global level) with a view of boosting their competitiveness, which is essential if they are to survive in today's value system, based on the only variable of instantaneous interest: very short-term financial profit, the foundation of capitalism. This context modifies the meaning given to the very notion of work and production, but at the same time shakes up rigid frameworks, inertial habits – in short, everyday life. It therefore creates disorder, sometimes creative, of desires associated with notable social progress, but also of risks, anxiety and a certain philosophy of resignation, because we have to adapt to technological evolutions that have not been co-constructed (Schumacher 1973a, 1973b; Puech 2008).

For Puech (2008), this situation is partly due to the fact that there is no real reappropriation of decisions by the public. He quotes Gandhi (1969): "I am absolutely convinced that no one loses his freedom except through his own weakness. However, this situation must take into account the fact that we live in a society with all the relations of explicit and, above all, implicit subjection that lead individuals to fit into 'socialization molds'" (Kaufmann 2001). This training (this molding) results in inertial forms of decision-making with the risk of cognitive due to cultural anchoring. According to Maalouf (1998), the context translates into the following reflections:

> Each of us has to make our way between the paths we are pushed into, and those we are forbidden to take or that are sown with pitfalls beneath our feet; we are not immediately ourselves, we are not content to 'become aware' of what we are; we become what we are.

At the same time, in line with Ulrich Beck (2003), we are moving from an industrial society for which the central objective was a certain form of wealth distribution to a society focused on another form of distribution, that is, of risk. On this basis, risk becomes a constitutive element of society. It is of a new nature, leading to a redefinition of social and political dynamics, based on a new, heterogeneous distribution of risk and the development of individualization. Under these conditions, according to Beck, "the recourse to scientific results for the socially binding definition of truth is becoming *more and more necessary,* but at the time same *less and less sufficient*".

Giddens (1994) takes a slightly different view, describing modernity as an ambivalent mix of comfort and anxiety, security and danger, and softening and hardening in the management of social conflict. He distinguishes three discontinuities between modern and traditional social institutions: the speed of change, the scope of change and the intrinsic nature of modern institutions. Others could be mentioned, such as the irreversibility of transformations... These elements, which it is not possible to go into here, place the business and society in what we call postmodernity or modernity pushed to its limits. Giddens (1994) writes on this subject:

> The desire to better control our environment through ever-greater knowledge of it constantly leads us to question knowledge and to act on our environment by modifying it. The point is not that there is no stable social world to know, but that knowledge of that world contributes to its unstable or mutable character[3].

Environmental research (in the broadest sense) explores this unprecedented change in context in a very indirect way. Mechanisms for dissociating time and space, such as the relocation of social systems and the reflexive (re)organization of social relations, contribute to the dynamism and excitement of modernity. They lead to a suspicious, uneasy reality.

However, dependent on the state or on social security systems, the institutes in charge of expertise on risks (their research is a means of maintaining internal competence and of exchange with the places where new knowledge is created) are confronted with maintaining trust between those who bear the risks and those who produce them based on success stories that make it possible to highlight effectively measurable improvements. This means that, essentially, the search for cause–effect relationships is, from the outset, a practically necessary condition for the success of the action by leading to credible rational prevention solutions, because they are understandable by all stakeholders. Thus, in the recent past in the prevention of

3 Author's translation from the French version of the text.

professional and environmental risks, the study of the harmful consequences, in terms of accidents or occupational diseases, has made it possible to seek the causes, and, consequently, to reduce the undesirable effects (values exposure limits, standardization, regulations, etc.). Philip (2007) has defined the principles that apply to sustainable development, but this context is flawed in many situations (time lag, multi-factoriality, multiple origins, irreversibilities, failure to take populations into account concerned, etc.).

These different elements illustrate the fact that knowledge and understanding chronic risks prevention, even accidents, are at the heart of our "world" only mediated by a social existence (a recognition): it is not the scientific knowledge itself which must modify the living conditions of citizens, taken in the broad sense, but knowledge taken in forms of appropriation and specific enhancements which come under social debate. Scientific uncertainty is therefore increasingly reflected in a kind of demand for a representation that conforms to our own perception (with all of the limits we know) (Libaert 2003).

The society of risk, as developed by Beck (2003) and more recently by Nowotny et al. (2003, 2005) and Gibbons et al. (1994), introduces the need to recontextualize scientific activity in the field of a society that is both better and worse informed, concerned about an uncertain future and marked by environmental issues (including those related to work). We thus talk about responding to "social demand"; however, this demand is rarely expressed, hence, in some documents, the concept of "latent" social demand, which cannot be "average". It would be better to speak of research issues that concern certain social groups, but, under these conditions, we would lose the whole "unifying" aspect of the concept of social demand, a real rhetorical leitmotiv today...

NOTE.– (Responsible) relationship with the public.

Table 1.1, taken from Tait (2017), illustrates the difficulties related to the relationship between science and society.

The precautionary principle, which has been incorporated into the Constitution in a context of complexity and scientific uncertainty (André 2005, 2006). However, social pressure and the use of experts outside the strict scope of their knowledge raise serious doubts about the robust use of science in expert appraisal, if only because the "teachers" may be the originators of what may be questionable (André and Masse 2002)! This can lead to a desire to control science by a public which, whether conditioned or not, is more inclined to focus on the negative and irrational side of the risk issue than on the possibility of an approach based on a medium- or even long-term dynamic of the "cost–benefit" or "interest-disadvantage" type.

Group thinking	The opinions of small groups can be influenced by participants with strong opinions or those who lead the engagement (so-called dress effect)
Framing of questions	Unengaged citizens do not want to participate in hypothetical discussions – recruitment risks being biased in favor of those with a scientific background
Recruitment bias	Unengaged citizens do not want to participate in hypothetical discussions – recruitment risks being biased in favor of those with a scientific background
Conflicts	Why would there be agreement on an issue related to disruptive technologies, especially if there is a polarization of opinion
Disorientation	Some vast and multidimensional subjects are too general to have an informed opinion upstream
Weariness	In order to judge you have to look more deeply, which requires time and resources, insufficient for engaging on each question
Changing opinions	People do not yet have a clear opinion and will change their minds in the relatively short-term (faster than the duration of a research study)

Table 1.1. *Difficult relationships between science and the public*

According to Morin (2000), this situation stems from the fact that "we teach certainties, but never fundamental uncertainty". But is this possible? Julliard (2008) points out, in this relatively new context, that "the democracy by default of the past is being replaced by the democracy by excess of the present, where the citizen's problem is not to be informed, but to resist information". Today, there is a plethora of information channels, from radio and television to the Internet, blogs, forums, newspapers and so on. According to this author, we are somewhere between an "opinion" and "opinionated opinion", illustrating the fact that public opinion is an artifact... This view is backed up by Sfez (1992), who considers that "consumers of communicational products are supposed to be enlightened, in all matters of opinion, even though opinion is 'preformed', powerless because technology has become our living environment".

This well-known observation of a "fuzzy" and labile opinion, as in the case of genetically modified organisms or "human augmentation", recalls the "doxocracy" dear to Bourdieu (1973) and leads to reflection on a so-called science mode 2, carried out in concert with society, in connection with the social dynamics at work (Nowotny 2003, 2005). According to Blondiaux (2008), "the idea of an inevitable strengthening of citizen participation in political decision-making is also gaining ground every day in most major Western democracies". The questioning of the delegates of representative democracy therefore strengthens the power of public opinion, sometimes that of the street, supported by those disappointed with the power in place, whatever it may be. These rather new conditions must – at least in part – take researchers out of their discipline (Leydesdorff and Etzkowitz 2000), the "so-called" mode 2, which is an important political choice, would not be so new (see also Bensaude-Vincent (2009)). It is therefore a question of defining modes of action where science and society work together, avoiding, as far as possible, the unconsidered loss of confidence in the positive contributions of science. Mode 1, stemming from the end of the last world war, is an intellectual and lobbying construct aimed at justifying scientific autonomy (Bush 1980). For Foray (2001), this is an evolution where mode 1, which makes the assumption that scientists do not have strictly the same interests as society, would be subject to mode 2. As this subjection is still to be made (see also Zweiacker (2007)).

There must therefore be ways of "mobilizing" (but by whom?) this knowledge with a view to a form of social utility that must be perceived as such. Indeed, if we assume that it is possible to seriously anticipate a risk, and that reliable solutions are envisaged right from the design stage of an operation, the net result is that the risk will never be perceived! With significant heterogeneity in citizens' views of what is right and just, and with the unprecedented development of individualism (Bourg and Schlegel 2001), social demand is only present on situations deemed alarming by groups formed around this single problem, that is, when it is almost too late. Gauchet (2002) reminds us that:

> it's worth asking whether we're not witnessing one of the great oscillations that Western societies have periodically experienced since the 'invention of the individual'. Oscillations linked to the difficulty, if not the impossibility, of thinking together individual and society, to reconstruct society from individuals.

The demand for change is very largely based on real effects or impressions that affect individuals. As Gauchet (2002) puts it, the source of these problems lies in the fact that "distance and mistrust are the two sources of contemporary individualism".

Sloterdijk (2000), followed by Fleury (2002), introduces principles of disruption that force correction. Fleury writes: "To question the non-real (which is the accident) is to question the nature of the real that confronts us, whether its modality is visible [...] or invisible". Accidents, like pain, appear to be the privileged occasion for an encounter with truth. In any case, these disruptions induce a loss of confidence between science and society. It is on this basis that research is called for, with strong pressure to produce preventive solutions quickly: banning certain products, exposure limit values, collective and individual protective equipment, process changes, etc. It is a classic case of getting a (slightly) seized-up system up and running again, by greasing it up or fixing it! However, this means that the context is part of a rather acceptable rational dynamic thanks to the processes induced by the mono-disciplinary scientific approach. There is therefore no fundamental questioning of practices.

The principle of nonindependent research (involving at least a state-paid salary) necessarily implies forms of bi-univocal relationship between academic research and society. According to Julliard (2008), there are three instances of expression of the popular "will": democratic suffrage, opinion and the field or street... Until recently, it was the democratic elites, in conjunction with the power of scientists and companies, who promoted science to its current level. Various "affairs" have, at least in part, removed some of those who legitimately represented opinions to the benefit of players who are undoubtedly more militant. Recognition of the diminishing legitimacy of political representation is partially affirmed in the Lisbon Treaty on the European Union (2007) (quoted by Blondiaux (2008)), which encourages "an open, transparent and regular dialogue with representative associations and civil society" (article 8B).

However, this is no more than an observation: in certain situations, whether critical or rendered critical, the center of gravity of power has shifted from rationality to the immediacy of manifestation. Who, then, robustly represents today's opinion? Who can the scientist trust? In the previous period, as Bruner (2002) reminds us, "we might not unreasonably think that the science of statistics was invented to protect us from our great weakness in predicting. The fact remains that it is our hopes and desires that drive us". The stability of modes of thought and action in research is, to a large extent, linked to the existence of scientific disciplines (but how else can we do it!) and, no doubt for reasons of power, in the broadest sense, by the existence of plans: the Brussels Framework Programme for Research and Development, the European Research Area, and development framework programs, as well as the programs of major research organizations, ministries and funding agencies (ANR in France). These frameworks, which are far removed from new forms of opinion, are linked to the money made available through intentional incentives, and thus, for researchers, to the possibility of acting (more) serenely, at least for a time, in the areas funded...

Over time, this context leads to a strengthening of the mono-disciplinary culture and influences the societal fields in which states "invest" their best brains (or even those of others) (Hacking 2001). However, in the current scientific intellectual movement, whatever it may be, we can consider that the values of objectivity, coherence, rationality and justification are well respected (Berthelot et al. 2005). A number of elements need to be taken into account in these different areas:

– There is an "emergent" demand (we can assume that an anticipation work has been carried out?) which induces a more or less disciplinary parceling out of actions.

– This demand translates into a quest for performance for which, in the current time pressure to focus on an optimum number (i.e. the smallest possible number) of influencing parameters to improve/innovate for performance. As a result, lower priority elements, such as those corresponding to risks, are referred to other players, to regulations, for lack of culture, time or even lack of interest or simple understanding.

Under these conditions, it seems difficult to envisage an easy integration of all of the components needed to take into account the concept of risk avoidance (some of which are not even imagined by those who may be behind them) in the quest for the fastest possible economic performance. What is more, for society as a whole, media analysis rarely raises the classic question of the relationship between capital and labor, between the production of attractive goods and potential risks of technical progress, the birth rate, "eternal youth", health, pensions, immigration and intercultural relations, and so on. Where is prevention if there is no effective social pressure, even "at the café du commerce"? It is in this context that ecology has a hard time getting its message across!

These considerations illustrate the wide gap between the desire to approach a society, which is irreversibly being built up on a global scale, and the atomization of the work of the researcher, who has little or no mastery of the organization for which they exercise their talents: they do not think of themselves among the others and in terms of the community. Generally speaking, they are unable to distance themselves from others and ensure that they are different or complementary. "It goes without saying that science is a historically situated social activity and must be understood in relation to the contexts in which it exists" (Shapin 1998). The result is a certain form of de facto formatting, or even "embrigadement" of scientific development (Thuillier 1971). In the field of science alone, this formatting is defined by Kuhn (1983) on the basis of a paradigm: "Universally recognized scientific discoveries which, for a time, provide a community of researchers with standard problems and solutions" (Kuhn 1983; see also Omnès 1994).

Thus,

> to speak of a paradigm is both to say that there is a social constraint to think in accordance with a major example of explanation, the one that passes for particularly luminous at the time, and, at the same time, that those who do not think in this way, or who are too concerned with what the preferred mode of explanation does not explain, are momentarily marginalized (Descombes 1989)!

In fact, science and society are underpinned by two phenomena (at least): complexity and interdependence, both of which can be addressed through interdisciplinarity. Complexity is synonymous with a multiplicity of possibilities. To act, we need to reduce this multiplicity. This is precisely what trust allows (in principle) (Quéré 2001). This trust, which presupposes agreement on the loyalty of all actors investing in uncertainty, is only partial, for a variety of reasons (André 2006). For example, the Canadian Food Inspection Agency (CFIA 2008) conducted a study on the factors influencing public attitudes in the 21st century. The following elements are cited as:

– distrust of all institutions;

– dependence on advanced technologies;

– realization of the interconnectedness of actions and consequences – global scale;

– use and complexity of technology in human society;

– understanding the contribution of human choices and behavior in the face of danger;

– distrust of risk assessment.

At the same time, according to Coote and Franklin (1999), the public, in the eyes of political scientists:

– is unable to understand complex issues;

– is unable to form relevant points of view;

– believes everything they read in the papers;

– forms an opinion based on narrow, selfish concerns;

– is apathetic;

– does not want to take the time or make the effort to think (André 2006) about anything that does not directly affect them.

In short, all is well until the next crisis... Against this backdrop, Fischhoff (1998) presents the necessary elements of risk communication to the public in a presumably cynical manner. All we have to do is:

– get the numbers right;

– tell them the numbers;

– explain what we mean by the numbers;

– show them that they have accepted similar risks in the past;

– that it is a good deal for them;

– treat them nicely;

– make them partners;

– all of the above.

In principle, this general framework reaches the researcher who is not confronted with an interdisciplinary approach (on average, but there are always deviants). The aim of this mode of action is to translate "real-world" problems into scientifically possible interdisciplinary actions, based on a technoscientific evolution which, while often in continuity with previous activities, is not a prolongation of them (particularly in the associated social significance) (Gras 2003). While a large part of the industry has mastered these approaches for its own survival, and apart from the incantatory rhetoric of many principals, this desire to promote the idea that "the whole is greater than the sum of its parts" is being severely undermined:

– recognition by their peers difficult for scientists acting in an interdisciplinary;

– career problems (Vinck 1995);

– difficulty integrating the various disciplinary elements;

– how can we find out about the methods and practices of other disciplines that are necessary for action? How can we understand each other?;

– the "trial and error" approach implies knowledge of the dynamic of the various players involved, and does not fit in well with action programs;

– creation of new disciplines through mergers, bi-disciplinaries (e.g. biophysics, etc.) which restructure action within a new framework but using the same forms of "recognition" (adapted scientific journals, peers in the field, etc.);

– from an epistemological point of view, the creation of new knowledge does not follow a linear process. It is therefore difficult to synchronize knowledge from different disciplines.

In addition, the development of quantitative methods for evaluating researchers through so-called criteria, imposed by the massification of research and the immediacy of everything, results in a modest recognition of innovation, particularly if it is interdisciplinary in origin (see Chapter 1 of Volume 1). By attempting to reconcile the points of view between science and society, we are generally a long way off the mark... There is a risk of transferring responsibility for risk-related work to a few "scapegoats", because that is what they are supposed to do... So, if we are to make any headway in the field of risk, in a world of ever-increasing complexity, we need to think in terms of different modes of action. In this new adventure, made necessary, several operations must be attempted, relying on a plurality of players acting in a concerted manner:

– "visionaries" involved in forward-looking thinking, in contact with the world of innovation, research in the human and social sciences, etc. Their task is to define scenarios for debate with society;

– integrators responsible for taking into account, in anticipation, society's problems to be addressed through interdisciplinarity;

– the players, true project managers, have the difficult task of defining a methodology evolving through the use of experience feedback, and explaining to society how it is possible to "modestly" manage a certain form of complexity;

– disciplinary research players.

1.2.3. *Scapegoating*

> In a political, social and consumer vacuum, the formula for determining the likelihood of job replacement would be simple: if the technology available could match human behaviour and operate at a lower cost-per-production ratio, the job would be automated (UNESCO 2021).

> The 'trivialization of evil' is only effective because humanity as a whole sacralizes technology, and this sacralization takes place in a totally unconscious way (Decarsin 2021).

> The pandemic allows us to see how the health political, economic and technical systems cooperate to this end. To the detriment of freedom [...] and the global ecosystem, i.e. life, in an infantilism that prevents us from giving up our polluting recreational vehicles. Here's a totalitarianism we're increasingly powerless to resist because, since it's universal, it's also increasingly difficult for us to become aware of it and put up a resistance worthy of the name (Dufresne 2021).

For a long time, scientists were able to steer clear of crises associated linked to the emergence of risks considered unacceptable by society. Recently, however, society has begun to call to account those who may be at the root of new problems. In France, for example, the *Commission nationale du débat public* (French National Commission for Public Debate) is conducting debates on the useful effects and/or environmental and social impacts of nanotechnologies, or on the burial of nuclear waste (Métais et al. 2014).

Within this new framework, not only industrialists but also researchers from the public sector are invited to take part in the debate. Pro-foundational crises can probably be avoided by encouraging such debates between science, technology and society. The French government has wisely set up "professional" risk-management interface providers (INERIS, IRSN, INRS, etc.). The following pages reflect the author's vision of the role of these institutes in harmonizing the relationship between scientific knowledge of risks and hazards, on the one hand, and social representations of these risks, on the other hand. The existence of disagreements between the various stakeholders (which do not yet concern the world of academic research) result in crises and losses of trust (Redlingshoefer et al. 2014). Recently, this is what we experienced with the Covid-19 crisis.

The crisis, for Heiderich (2007), results from the combination of an objective event and a set of individual perceptions which can lead, through the associated tensions, to a symbolic disconnection. There are generally what he calls "media iterations" produced in the moment and "contingent by retinal thought", which are associated with the demand for a conforming representation, more or less linked to a prioris and cultural references (Libaert 2003). Research actors are only rarely prepared for crisis management, nor knowledge of the fragility of the systems in which they work, which can be of particular concern for the maintenance of trust between their world and society. Thus,

> it is not only the relationship with the other, insofar as he is different from his origins, which signals the crisis, it is also the relationship with the other, insofar as he is similar. Similar, but in its individual singularity: like me, an autonomous individual, therefore different from me. Incivility is also practiced between equals (Bensaïd 1999).

In one of his books, Girard (1982) examines the various stereotypes of crises linked to misunderstandings between stakeholders and "associated persecutions": it is the juxtaposition of several stereotypes in a single document that leads to the conclusion of persecution. Not all stereotypes have to be present. Three of them are enough, and often even two. Their presence leads us to assert that:

> – the violence is real;
>
> – the crisis is real;
>
> – the victims are chosen not by virtue of the crimes attributed to them, but by virtue of their "victim" signs, anything that suggests their guilty affinity with the crisis;
>
> – the point of the operation is to shift the responsibility of this crisis and to act upon it by destroying said victims, or at least expelling them from the community they pollute.

For Girard, this model is universal, both when reading historical texts and on a planetary scale. If this myth is found in a stable way in time and in space, the French National Institute for Research and Security (*Institut national de recherché et de sécurité*, INRS), one of the components of the Social Security Prevention Institution, administered by the social partners, but financed by the state, was the object a few years ago of a keen and critical gaze from the media, then from the state, associated with a form of "moralizing one-upmanship which seeks to flush out misguided goodwill or help bad at hiding" (Slama 1993). This institute of research and scientific expertise can be one of the examples of challenges by society and can serve as an element of reflection (for tomorrow?) for the world of research and expertise associated with a context of social utility...

What "crises" for the research and expertise center (which the author knew well) of the social insurer responsible for occupational risks! Asbestos, glycol ethers, possible links between genetics and work, the possible toxicity of aluminum, etc. Today, nanoparticles are emerging in a context where asbestos has rightly left "disastrous memories". How will the "ordinary" research community, committed to "sensitive" subjects and less trained in expertise, resist crises linked to risks in the future?

In this example, can we consider that there is a clear desire to harm the company in the workplace on the part of the social insurer? Or on the contrary, the emergence collective worries reflected in the persecution of the right "scapegoat", the right "target"(!)? This is a fundamental question that needs to be considered! Could this institute have broken a certain "harmony" that led to this desire to purge in the name of a false "social correctness"? Has there been too rapid a change in the social acceptability the nature of risk changed? These different elements focus the questioning on what represents acceptable (professional) risk in today's rapidly changing world, but also on why the emergence of media phenomena, true sacrificial preparation, ultimately seeking to make the victim alien to their basic mission? Based on the victim stereotypes presented by Girard (1982, 1973), we can show below that this institute has all the "assets" it needs to achieve a form of persecution aimed at opprobrium and elimination by militant actors.

1.2.3.1. *The reality of violence*

Both technical and organizational developments make it impossible to treat the world (including the world of work) as an idyll between employee and employer, or between citizen and state (see Gilets jaunes) because of the high human and social costs. Almost permanent technical and organizational change, precariousness, unemployment and globalization are all factors that increase citizens' uncertainty about the world of work, which also provides social status, meaning in life, etc. (Roman 1998). In this context, where there is (alas) no such thing as zero risk, there are operators who suffer from accidents or illness (the approximate cost is around 10 billion euros a year for France (paid by social insurance), but this probably does not take into account all of the factors associated with the occurrence of the accident or illness).

If employees are, on average, increasingly removed from risk areas in production workshops thanks to the automation of operations, we expect – as has been observed for several decades – a certain reduction in the number of work accidents. Today, however, the hundreds of fatalities every year show that further progress still needs to be made, always with the aim of making work less dangerous. It has to be said that the economy is largely shifting from a production-based economy to one based on knowledge and associated services, systems for which the rules of operation have not yet been fully stabilized, but which result in fewer accidents.

However, at the same time, the number of occupational illnesses is on the rise, from musculoskeletal disorders to more serious, long-lasting occupational cancers. Against this backdrop, asbestos continues to focus the harmful effects of work on insufficiently informed and protected operators. Thus, violence is real, even if progress has been made and its effects continue to benefit the world of work. The same is true, albeit to a lesser extent (because it is not financed), for citizens suffering the effects of various forms of pollution, for example.

1.2.3.2. *Crisis reality*

Godard et al. (2002) illustrate the existence of organizational continuities in the *Traité des nouveaux risques*, particularly with regard to risk expertise and management. They write:

> Does the insistence on scientific risk assessment in the context of precautionary policies appear suspect [...]? It is said to perpetuate the scientistic myths of the preceding historical period, or to convey strategies aimed, under the pretext of rigor and objectivity, at evacuating legitimate social approaches.

By relegating those who are affected to the role of suffering victim and by reinforcing the primacy of scientific advice, we create tensions, even the emergence of crises. Audoin (1999) recalls that "our body has become our place of deployment. We maintain it because it is our intangible capital". Thus, because of the development of self-care and the appearances of what we want (or seem) to be, there is an evolution in the social acceptability of risk in the context of work (Demortain 1999) and therefore to be an effective partner in dealing with the problem. In this context, the taste for individual risk is offset by the aversion to imposed risk. Any information likely to associate a new risk with a possible effect on someone's body then translates into new anxiety, anxiety reinforced by the media which "fill with superficial information what they ignore, this real complexity that 'no scholar masters'" (Aubens and Benasayag 1999).

Indeed, this extension of a desire for social control is the result of crises linked to the emergence situations: the Chernobyl cloud, AIDS and contaminated blood, dioxins, mad cow disease, GMOs, Covid-19, etc., leading to a loss of confidence of society in scientific experts and in the powers that govern them (even if experts are required to be independent). There is, on the part of the public, suspicion and a desire for moral superiority, even though generosity is always present in its actions in principle. Indeed, there is normally action in the name of an ethics of the good. However, it is possible to agree with Slama (1993) when he writes that "the intrusion of associative militancy into scientific territory takes on the appearance of an inquisition. It's a propaganda operation from which all oral criticism in the scientific field is eliminated". It is also important to bear in mind that, in the field of occupational hazards, which are often multi-faceted in origin, the responses of scientists and experts in the field to the public's demand for information are, for a variety of reasons, generally inadequate.

Indeed, for some scientists, the complexity of work may be an obstacle that can be circumvented by simplifying situations or, at best, by reducing them to the analysis of their elementary components (this is a classic return to our mono-disciplinary knowledge, effective but generally insufficient at providing an answer that takes all factors into account). The scientist then explores what they have understood about the issue, providing society as a whole with information that is normally honest, but partial and even truncated in its relationship with reality. In fact, there is a growing gap between the satisfaction of social demand, complex and immediate, and scientific responses that are marked by uncertainty and time consuming (André 2000), or even a translation into reductionism that is inoperable for the public through the use of forms of expertise.

This lack of understanding of the gaps between validated and robust knowledge, on the one hand, and the aspirations of the social body, on the other hand, is easily transformed into a trial of intent, combining nominalism, amalgam and Manichaeism (Slama 1993). The imperfect mastery of the links between accidents, on the one hand, and diseases, on the other hand, and working conditions induces, on the part of the public, a desire for social control tending toward persecution. It seeks action but does not have the means to act on the actual causes, hence the specific sensitivity to the search for pseudo-causes that it understands, as permitted by its culture. This is why they are so sensitive to the search for pseudo-causes that their culture allows them to understand, enabling them to express their discontent in a variety of ways. "Members of the crowd are always potential persecutors, because they dream of purging the community of the impure elements that corrupt it, of the traitors that subvert it" (Girard 1982). So the crisis is real:

– The victims are chosen not by virtue of the crimes attributed to them, but by virtue of their signs of victimhood, anything that suggests their culpable affinity with the crisis.

– First and foremost, the social insurer's and research center has a duty to provide scientifically and technically validated information on changes in the world of work: changes in technology, work organization, and the complex relationships between finance, the employer, employee, social insurer, state, Europe and the world.

– This “outside” knowledge is necessary to invest in the field of occupational risk prevention, inducing the development or adaptation of knowledge on measurements (allowing to check whether the regulations are followed), as well as (and rather above all) prevention engineering. This last strong component allows the advice of business actors with a view to improving working conditions within a socioeconomic framework and imposed standards.

In addition to these natural axes, work focused on occupational health has been developed, allowing the search for links between physical, chemical, psychological, organizational difficulties, etc., and the appearance of occupational diseases. This research was made possible by the knowledge that the institute (or its Western counterparts) has of the world of work and by the existence of structures acting internally in toxicology and epidemiology (and also perhaps because of the numerical weakness of researchers from national research organizations (NROs) acting on this theme of public utility which only imperfectly responds to the logic of scientific in-depth study and technological development enabled by research in engineering sciences).

Given all of the difficulties, even if all of the organization’s forces were concentrated on the occupational health component (which, however, corresponds to less than a few thousandths of French research), it would be impossible for the research center to provide the minimum satisfaction to all social demands (numerical weaknesses and the need to anticipate emerging problems). This situation can be found in most developed countries. Even if this is a mission that complements the basic, information-oriented mission of the social insurer, it can be highlighted in the media because of the concerns and uncertainties associated with illness.

However, in the national context, there aren’t any (or are few) organizations likely to cover this field of action. It is therefore (still) the INRS that is often considered, by those outside the social security as one of the main centers responsible for the development of knowledge in the field of occupational health. Even if, in France, these missions are now shared with national structures, it is sometimes assigned intentions that it does not have, in the name of a certain “socially correct”, non-virtuous attitude, supported however by proven facts: accidents and illnesses.

And yet, its mission is to inform, in a way, to demystify: “It’s not a matter of deciding for others, but of enabling them to do so, by their own means, by providing them with the elements enabling them to judge” (Revel 1988). In fact, this institute is given far too much scientific power and scope for action than is reasonable! The second element in the answer to the question, centered on the guilty affinity with the crisis, is linked to the specific nature of this organization’s parity management. It is, in fact, administered by a genuinely joint board (unlike the other components of social security: sickness, old age, family allowances). Thus, the reading of newspapers, following investigations by the General Inspectorate of Social Affairs (Desriaux 1999), aimed in principle to the study of possible links between INRS and the Institut de veille sanitaire (now part of ANSES), resulted in some powerful images of possible but effective employer control! The media's accusation is considered real by many, and naturally induces the bias necessary to reinforce suspicion. Finally, last but not least, money! If the institute is to be able to provide serious (and therefore validated) information to all corporate partners, it must have the means to achieve its objectives. It must therefore have sufficient funds to carry out its mission, without having to seek out private or other partnerships that may distract it from its public service activities. For some in the media, the institute's "riches" are real, but naturally misused, because they are linked to misdirected activities. The trial of intent is not far off. INRS thus has the third label, through the presence of a certain number of victim signs.

The aim of the operation is to shift the responsibility of this crisis and to act on it by destroying the victims, or at least expelling them from the community they "pollute". By classifying the organization as de facto controlled by the employer, marginalization necessarily follows by casting doubt on its "neutrality" or "transparency". But there are periods of "social peace" between crises. Moreover, many experts sometimes rightly consider that privately funded scientific publications are, on average, less "neutral" than those financed by the state. Without wishing to enter into this difficult debate, as each situation is a case in point, there are clear cases of "orientation" in writing (though perhaps not unilaterally), some of which have been the subject of legal settlements (Malka and Gregori 2005). According to Girard (1982), "ritual thought seeks to sacrifice a victim as similar as possible to the monstrous double. The marginal categories from which sacrificial victims are frequently recruited do not perfectly meet this requirement, but they are the least bad approximation". According to this author, the victim must naturally be made a stranger, isolated from others by limiting contact with "healthy" members of the community – in short, marginalized. This dialectic aims to channel forces within the limits by imposing ritual remedies. Does this not sound a little like Oedipus trying to kill his father?

If, in this situation, the public transfers its anxiety and to the scapegoat, the latter always loses out in the game of after-the-fact communication. Indeed, it is always difficult to lateralize a real problem by asking others to look at their own responsibility. It's also tricky to be able to justify yourself on every point. Indeed, in any human endeavor, there are always flaws. For the institute too, there is no such thing as zero risk, even if there is transparency in our activities (and if there had not been, would the public have been aware of the crises mentioned in this document?) So, the right of reply is difficult, and it is a delicate game, well known to communication specialists, a game they wish to avoid. When a crisis arises, how can we try to communicate on a conflictual subject by purging the discourse of its passionate character and reducing it to its rational content (when it exists, insofar as we often find ourselves in conditions where the effects of innovations are not really under control)? In fact, this is practically out of the question!

In the social issues discussed in this chapter, given the complexity of information and regulatory systems, all of the players involved have their share of responsibility for the choices made then and now. Progress is (and always will be) necessary. In this way, the sacrificial myth is created as a "reaction to the decrepitude of behavior, the resurgence of consciences confronted with the spiral of irresponsibility" (Lipovetsky 1992). How, in this context, can we believe that generous and ideal proclamations and virtuous protests, will, with a vision of exclusion, make it possible, in particular at INRS, to always be up to the challenges that it must overcome in its mission of knowledge and prevention of occupational risks? This organization has therefore all four "qualities" of a scapegoat! These "cure, of course, neither real epidemics, nor droughts, nor floods. The crisis affects human relations inducing a process of bad reciprocity" (Girard 1982). External suspicion turns internally:

> Whatever their true causes, the crises that trigger the great collective persecutions are always experienced in more or less the same way by those who suffer them. The most vivid impression is invariably that of a radical loss of the social itself, the end of the rules and differences that define cultural orders (Girard 1982).

The constraints induced by "persecution" should naturally be passed on to the "victims" who unite against them by virtue of their belonging to minorities. If certain elements of the institute are attacked, other links in the system are left in the shadows, which may characterize a certain internal disassociation, reinforcing individualism and a certain desire to belong to "normality". Argyris (1999) shows that individuals often adopt defensive routines that insulate their mental models from any challenge (see also Heiderich 2007). It is difficult, however, to be both a member of the in-house team and involved in the destabilizing actions mentioned in this chapter. But this is all part of life...

One possible explanation for this situation is that all of the state-owned organizations, which avoid the "employer" bias (even if they have industrial contracts), are still in a position of relational stability and trust on the part of society. In principle, therefore, public authorities provide citizens with an element of operational reliability, even for research organizations such as CNRS, INSERM, INERIS and CEA, which are increasingly obliged to seek private funding in order to operate.

And yet, this same public is virtually excluded from the decision-making process! Clearly, "experts are unable, given the current state of knowledge, to provide confident answers to the questions posed, or to recommend measures to restore confidence" (Roqueplo 1996). This phrase is associated with the uncertainty of the scientist in relation to the complex issues of risk or vulnerability; it distances them from the dynamics of decision-making which cannot be directly expressed through the mastery of knowledge alone. Yet, as the same author reminds us, there is "an obligation, despite the inadequacy of scientists' knowledge, to provide the 'causal knowledge' required by policy-makers". In principle, there can be no identity between the scientist who defines knowledge and the expert who organizes it with a view to an end, which corresponds to the work of experts and deontological regulatory mechanisms (Lipovetsky 1992).

Slama (1993) points out that "the progression of social control measures to the detriment of individual autonomy and the transfer of democratic legitimacy to experts are slow, subtle phenomena, due to a necessity that is not perceived as ideological". It is this concept of expertise, which is insufficiently communicated, allowing for the alert is one of the criticisms sometimes leveled at INRS. Once again, there is an amalgam. In addition, Etchegoyen (1999) has identified six "perversions" in law and morality that "reveal flagrant contradictions between the two concepts of responsibility legal and moral responsibility":

– The desire to do nothing: legislation acts as a brake on risk, limiting innovation. "It is always more difficult to prove the responsibility for inaction than for action".

– Fear of punishment: "we forget the other in order not to forget ourselves. This fear in fact aims to suppress the moral concept of duty".

– Moral oblivion: "victims can't stand the fact that the perpetrators don't feel responsible or don't admit any moral responsibility"; compensation is not enough! Criminal law is being abused, illustrating the lack of moral responsibility.

– The impossibility of thinking about accidents: "we no longer support the idea of damage without those responsible who would be there to pay in every sense of the word, from revenge to compensation". This is one of the areas where we must invest in prevention to help, in particular, from the design of the processes, the partners of the companies.

– Deflecting it onto abstract causalities: "there is a flagrant contradiction between personal (moral) responsibility and that transferred to the company" (example: consumption of tobacco, alcohol, psychotropic drugs, etc.), the problem of compensation being outside of this concern.

– The obsession with definition: "you have to define everything to cover yourself for fear of punishment".

These various elements, highlighted in Etchegoyen (1999) and quickly recalled, are precisely the subject of the search for a redefinition that is undoubtedly very provisional in the social acceptability of risks. Multifactorial, long-term risks, which may affect offspring, are a source of concern:

> We know what place the delicate associations of falsity and sincerity occupy in our psychic activity; the need to believe, stronger than the desire to know; bad faith, by which we take the precaution of concealing the truth from ourselves to be more sure of our firmness, when we deny it to others; the reluctance to acknowledge an error, unless we can attribute it to our qualities; and last but not least, our ability to implant in our minds those systematized explanations of reality we call ideologies, a kind of machine for drawing out facts favorable to our convictions and rejecting the others (Revel 1988).

For other players in society-related research, this may be an opportunity to reaffirm their independence in judgment. By making it more visible to society, there is a way out of this vicious spiral. Or would it be necessary to practice a new form of holocaust, a true sacrificial act of expiation, so that the survivors can then act in a "socially correct" way, associated with the shame of the bad deed? This is what happened after Chernobyl (1986), in a different context! Thus, "modern insecurity does not consolidate anything, even by ricochet. Far from supporting it, it directly undermines social cohesion, sowing uncertainty everywhere. And what we see, when we look at the most insecure areas, is a stupefied, powerless community that begins to doubt the benefits of solidarity, in other words, of itself" (Peyrat 2003).

For the public, the research laboratories of large organizations are still adorned with a certain innocence, a humanism and a good taste of honesty, but the contractual relations which are today necessary for their dynamics can lead to certain unfortunate amalgams and, as a result, crises over a certain failure of the knowledge production system. Scientific articles are increasingly withdrawn (see Marchand (2022), in the case of the Covid-19 pandemic). The image of "beggars", which is often propagated, induces an effect of vassalization vis-à-vis the funder who, in return, is brought to bear the responsibility for his actions, but to what extent...

In fact, the list of irritating topics is growing: pollution of all kinds; suspicious burial of nuclear waste; nanotechnologies (Agora 2008; Sauvé 2008); convergence (André 2017); transhumanism (COMETS 2006); citizen control (Dutton 2004); Wi-Fi (Layet 2008); data files (see Edvige dossier withdrawn by the state, RF 2008), etc. In this new context of loss of public trust, it is not because we have expert organizations with the "right" victim symbols that public social agreement on academic research will be maintained in the long term. We therefore need to think quickly about better relations between science and society; this is part of the researcher's mission. We need to take into account (Baude 2006) simple, robust principles such as:

– not giving the impression that there are stereotyped methods applicable to every situation;

– committing only to scientific facts and avoiding "social correctness" (Godard 2007);

– respecting the "ordinary" functioning of research: autonomy, validation by peers;

– respecting laws and ethical principles.

1.3. Modes of action

> It's quite true that all science is the result of individual experience, but there is another domain which is not individual experience and which is of paramount importance: it is the domain of morality, including the appreciation of moral institutions. And there are traditions that are not based on individual understanding or insight, but on the experience of groups that were selected for certain practices that may be purely habitual and have never been fully understood (Hayek 2022).

> In a culture saturated with technologies for appropriating our attention, our interior mental lives are laid bare as a resource to be harvested by others (Crawford 2016).

> Once upon a time, a ploughman was sweating in his field a few leagues from Baghdad. There passed before him a hideous and disquieting man, dressed in black and with evil omens. 'Who are you, stranger? Where are you going? What are you doing?' asked the peasant. In a voice that was surprisingly gentle and kind, the dark being replied: 'I am the plague, and I'm on my way to Baghdad, where I'll kill 500 people'. Some time later, having educated himself on the ravages of the epidemic in the capital, the ploughman suddenly confronts the man in black on the road. 'Hey, liar! Why did you tell me you were going to kill 500 people in Baghdad, if over 5,000 deaths have been reported?'. In the same gentle tone, the plague replies: 'I didn't lie to you, and it's not my fault. I only killed 500. The others died of fear' (Vallayes 2020).

In modern social systems, greatly disturbed by the media coverage of everything, the very strong time pressures and the inclusion in procedures of avoidance of social reactions favored by scientific uncertainties, lead, as has been widely expressed, too often to truncated debates. It is true that the disciplinary functioning of research (working in "silos") does not easily allow complex problems of interest to society to be tackled and favors this situation.

Moreover, when challenged, researchers generally prefer to confine themselves to their own system (of values?), a system in which their social responsibility is (still) a moot point. This situation is largely adapted to the physical, chemical and engineering sciences, where innovation comes first, and other aspects are best considered in later stages and handled by other players (it is this nonintegrated form of sub-contracting that is one of the reasons for the existence of safety research institutes in most developed countries). As already mentioned, the situation is quite different when research directly concerns human beings, where integration the visible impact of life sciences knowledge on the health of citizens (see, for example, UNESCO 2005).

In fact, as we have seen in recent "cases", leaving the researcher in contact only with his stakeholders and principals in a system not controlled by society can lead to dangerous drifts. Lavelle (2007) considers that "it is up to ethics [...] as the expression of a social demand of reflection and regulation, the ambiguous mission of assuming the shortcomings of politics. Ethics is thus called upon, rightly or wrongly, as the form of alternative thought and action most likely to limit the oversights, damage and excesses of technology and, moreover, to uphold the requirements of democracy".

While this may be acceptable, putting it into practice in the everyday research is not so simple, for a number of reasons that have to do with the spatial and cultural proximity of the various players involved. At the same time, society's desire to control research too closely can lead to equally totalitarian drifts, as a result of control by militant actors, the bearers of "opinion" often in the minority, but expressing themselves on behalf of society as a whole, and by the perennial display of the precautionary principle (a barycentric shift influenced by media noise). While it may seem suicidal to support the confinement of researchers to their ivory towers (disciplinary ghettos), it is important to think about ways of avoiding a blockage in the production of knowledge, the effects of which society will soon see. However, experience shows that the public has little room for maneuver in this context, as it is now obliged to accept technologies and consumer goods that have been decided elsewhere. The aim of SRR, presented below, is all about bridging the gap between the complex and diverse aspirations of the social body and the estimated effects (evolving over time) of the transformations of the world brought about by invention and innovation.

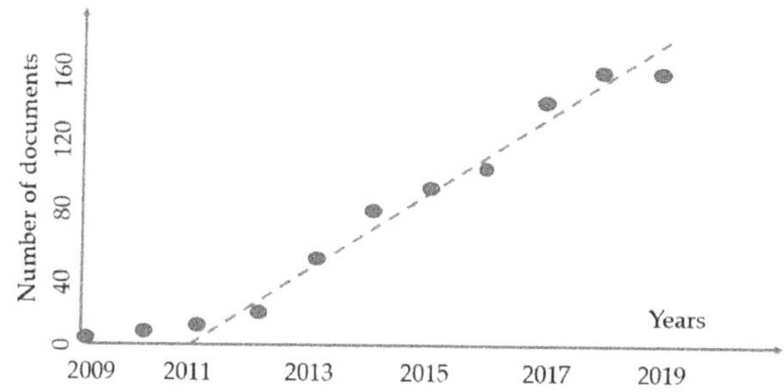

Figure 1.5. *Trend in the number of publications on the theme of research and innovation managers*

On this important topic, there is a wealth of work on conceptual thinking and empirical studies on the subject of innovation (Timmermans 2017). Moreover, this area has been promoted by the EU in its Europe 2020 strategy (EU 2019; Horizon Europe 2019) with the aim of making research and innovation more sustainable and inclusive. Figure 1.5, taken from Nazarko (2020), shows that this line of thinking has been developing over the last 10 years.

In essence, there is a great deal of asymmetry between specialists and laymen (Estrades and Rémy 2003; Cuppen et al. 2019), leading to misunderstandings with the creation of dissonance because the sequences of actions/behaviors can force us to readjust our positions before finding ourselves in a more coherent and ultimately pacified universe (Leroi-Gourhan 1965; Kaufmann 2008). The result may be the radical rejection of a new technology or a new product via the precautionary principle, which cannot be invalidated by the reductive authority of the "scientist", too confined to their own discipline, as they generally have no control over the consequences of the uncertainties associated with their inventive work especially if they are to be held accountable (Owen et al. 2013). According to Ramonet (1999), four cultures coexist in European countries: anthropological culture with its traditions and customs, humanistic culture (with its interest in Man, Nature, etc.), scientific culture ("knowledge for knowledge's sake") and mass culture, in which "information blurs into each other". According to this author, the connections between them are very weak. This chapter attempts to reconcile one form of scientific "angelism" with another form of social "barbarism", in an attempt to reduce the gap between the speed of "signs" and the slowness of the evolution of "meanings", and thus better justify the ends chosen from among all of the admissible ends.

To this end, without wishing to move to a system of "accounting" responsibility with relevant pseudo-indicators (see section 1.3.2), based strictly on explicit contractual provisions (overly precise ISO 9001 or ISO 14000 "checklist"-type approaches, for example), possibilities for reconciling points of view are presented below (Dreyer et al. 2017; Rask et al. 2018; Shin and Park 2019; Walecka-Jankowska and Zimmer 2019; Yun et al. 2019; Owen et al. 2020).

1.3.1. *The project framework: socially responsible research*

> "The legitimate use of contradiction," adds Simone Weil, "consists, when the human intelligence is faced with the necessity of accepting two incompatible truths, in recognizing them as such, and in making of them as it were the two arms of a pair of pincers, an instrument for entering indirectly into contact with the sphere of transcendent truth inaccessible to our intelligence" (Weil 1955).
>
> Schumpeter's cumulative innovation is the primary source of growth: every individual will tend to under-invest in innovation, because he does not internalize the improvement in knowledge that an innovation brings to society. Knowledge that an innovation brings to society and the fact that other innovators in the future will be able to build on this knowledge (Aghion et al. 2020).
>
> There are historical moments of indecision or 'bifurcation' in which we don't know how a community or socio-cultural society will continue (Rosa 2020).

It is possible to hypothesize that the so-called "pure" sciences of conceptual deepening are a long way from immediate application. However, this proposition does not stand up to analysis, as the definition of regularities has led to numerous technical advances that we will not go into here. There are therefore probably no sciences, and in particular those concerned with artifacts, that are disconnected/separable from the social context. This can be seen in the proposals they submit to calls for tender, where many researchers, no doubt because they are asked to do so, mention the (sometimes fanciful or totally unrealistic) spin-offs of their future work in terms of applications. As stated in section 1.1, it is not easy to define a Mode 2 research activity in relation to "popular expression", since opinion is neither structured nor stable over time. It seems potentially useful to have an independent organization whose mission would be to inform (non-instrumentalized) scientific choices by collective decision to try to take account of the remarks made in this chapter (but is this easy to achieve?).

Indeed, research cannot function in a permanent and random interaction of opinion, particularly in response to unsubstantiated reactions (whatever the field: scientific, social and/or political). In the absence of a structure such as the "Commission Nationale du Débat Public" (CNDP), which would engage in an in-depth discussion, the aim of socially responsible research is different. This assertion redetermines the researcher's responsibility, but in fact leads to a judgment a posteriori of their activity (contract of trust).

To this end, it could be proposed to award a label to research teams (unless this involves an internal cultural change, a form of self-certification) which, in addition to their scientific relevance and performance, propose to make a serious commitment to broadening their activity, with a view to reinforcing their position as a useful player in the social progress made possible by science. A non-exhaustive list is presented as follows:

– traceability: this investment is essential for building trust between partners, particularly with companies (quality approaches, best practices, etc.);

– responsibility compliance (*a minima*) of regulations research operations: protection of researchers and the environment;

– compliance with ethical laws (human experimentation, animal experimentation, data protection, etc.);

– originality/novelty: reflection on the launch of research operations whose outcome cannot be predicted from knowledge of the "prior art"; this means a better creation of knowledge, effectively new through scientific interdisciplinary projects. It means constantly questioning our own practices;

– conscious analysis of the uses of research results for society (this is a useful reflection, but difficult because of uses/transgressions of the same concept for different purposes);

– in-depth reflection on the uses of artifacts in society use and associated disruptions, recycling, risks of major irreversibility effects on the environment in the short, medium and long term; relations with mankind: risks, social justice and equity, development of the human personality;

– periodically revisit the various items in line with advances in scientific knowledge and their effects;

– mission to alert if new risk situations appear to be emerging.

In this way, following the example of CSR (corporate social responsibility), a research team/unit or group in charge of a pre-POC committed to "socially responsible research" in its operations, would do well to:

– comply with the ALARA (As Low As Reasonably Achievable) or general precautionary principles;

– undertake a risk analysis to (re)define appropriate protection methods, in a written document – a research team's operating charter – that can be communicated to the supervisory authority(ies). This approach gives teams a sense of responsibility and avoids the "crime of ignorance";

– inform the other laboratories/teams concerned (pooling of progress).

These various elements naturally need to be explored in greater depth, clarified and shared. They could lead laboratories engaged in these (new) practices to be awarded a possible "socially responsible research" (SRR) label. In this context, we need to analyze who would provide such a label, how the associated activity, what would be the benefits for research units linked to this label, how would we communicate with society on this important development. These elements are covered in sections 2.3 and 2.4.

NOTE.– CSR (corporate social responsibility).

CSR standards and practices are not imposed by law or treaty. They derive from the market, advocacy and investor expectations. They also require companies to behave in certain ways when operating abroad. While governments have encouraged the adoption of CSR standards, they have generally avoided legislating in this area as they have for environmental, social and governance standards (FI2020). Although they are geared toward risk disclosure, and therefore on a different plane to CSR, it is once again private-sector standards, independent of state regulation, that have a direct effect on corporate behavior.

According to Debref et al. (2019), this is a point of the ISO 26 000 standard. Its boundaries have expanded to include stakeholders (Donaldson and Preston 1995), to innovation processes, with upstream consideration of potential impacts on society (Pavie 2012), or by using social and environmental goals as a starting point for innovation (Ingham 2011; Temple et al. 2018). Innovation can then no longer be considered technological, as it must incorporate a principle of responsibility.

1.3.2. *Other considerations*

Several themes could be envisaged to operationalize the elements defined above. As this is a pilot action, this proposal for a possible charter could be reviewed periodically (if only in the light of regulatory changes and feedback). In fact, what we are looking for is a cultural change.

1.3.2.1. *Theme 1: aspects of "scientific performance"*

It would be useful to have a document defining the aims and means of the scientific research carried out in the pre-POC unit/group in relation to certain societal needs. It is a reflection document that defines what the group means by performance. It is only natural that there should be a gap between this reflective analysis and the traditional ways of defining this concept.

1.3.2.2. *Theme 2: legal aspects*

Compliance with the laws and regulations laid down by society through its representatives is a prerequisite for "socially responsible" research.

1.3.2.3. *Theme 3: management aspects of research under uncertainty*

In the absence of regulations, in terms of criminal liability, there is no liability without a text. However, if a case were to be brought before a judge in the event of exposure to risk or damage to others, the researcher's legal liability could be sought on two grounds (if the constituent elements of the offences are met) of the researcher could be sought on two grounds (if the constituent elements of the offences are met):

– criminal liability: the offence of deliberately endangering others (art. 223-1 of the French Penal Code) or a characterized fault that exposed others to a particularly serious risk that the perpetrator could not have been unaware of (art. 121-3 paragraph 4 of the French Penal Code);

– civil liability: the researcher has caused damage to others and may be held civilly liable.

NOTE.– In the current state of the file, there are no regulations specific to "uncertain risk" research, but the precautionary principle remains valid. It may be proposed that:

– The laboratory takes sole responsibility for an action involving uncertain risk and it relies on best practices validated by national or international agencies (this means that it has thought the matter through).

– The laboratory or pre-POC group questions its supervisory authorities about the validity of its wish to act in the field concerned. It is then up to the supervisory authorities, in conjunction with the group, to bring together experts who can either confirm the group's wish, or consider that responsibility should be taken at a higher level. At this stage, it is all about mutualizing thinking for "responsible" action.

1.3.2.4. *Theme 4: ethical and moral aspects*

Under no circumstances should responsibility be a tool that prevents science from pursuing its quest. We need to give priority to a plurality of viewpoints, in other words, to a more deliberative, more controversial approach to science – in short, new forms of deliberative rationalism, because they are collective. In other words, we need to learn how to construct a scientific object and problem with other experts. Organizing debates, led by the research group, can be a way of better linking science, technology and society. In the case of more controversial research, the researcher's responsibility must be redoubled. At the same time, all exceptional situations, and the reasons of state or emergency invoked, can create situations of lawlessness. Therefore, "getting out of ethics" must not be equated with getting out of the law, which would lead to "nonethical" being legally conditioned. Let us not forget that researchers have a redoubled social responsibility, precisely because they are scientists, "savants" and therefore, in principle, entitled to be "knowers" and experts.

These various elements must therefore be taken into consideration, so that, in consultation within the unit/pre-POC and with the supervisory authorities, unit charters can be used to experiment and pursue, "in real-life situations", the improvement of confidence that the company (or certain segments of it) to allow scientific adventure.

1.3.3. *Socially responsible research*

> If you think about it for a moment, you'll realize that we all produce to satisfy the needs of people we know nothing about, just as we all live thanks to the efforts of people we know nothing about. Which means that the current interaction producing the volume of goods that sustains the world's population, is the work of people directed for ends they know nothing about, and other people relying on other people who know nothing about them (Hayek 2022).
>
> Our morality is not a creation of our reason and cannot be the creation of our reason (Hume 1975).
>
> Contradiction, as Plato knew, is the sole instrument of developing thought. But there is a legitimate and an illegitimate use of contradiction. The illegitimate use consists in combining incompatible assertions as if they were compatible (Weil 1955).

In general, based on forward-looking thinking, there are numerous action programs published by various countries (the United States, Canada, European Union, United Kingdom, Germany, France, etc.), essentially aimed at research operations (programming). The industrial sector, on the other hand, requires rules of good practice for its risk control and liability management, which may be found in

ISO or standardization (CEN). In any case, the absence of reliable information in terms of possible risks, we need to apply the precautionary principle, currently enshrined in the French Constitution. In this context, we need to define the conditions for maximum protection of research operators, as well as those around them (environmental issues).

If this information can be easily communicated, it is then necessary to examine the situation on a case-by-case basis, in order to define the conditions for "minimal" risk research at local level. Following the example of CSR (Wikipedia 2022), the "socially responsible research" (SRR) label should/could be awarded to laboratories that:

– comply with the ALARA "or general precautionary principles";

– engage in a risk analysis to (re)define appropriate protection methods, which will be the subject of a written document, the operating charter of a certified (?) research team, which can be communicated to the supervisory authorities. This approach gives teams a sense of responsibility, and avoids the "crime of ignorance";

– inform the other laboratories concerned (pooling of progress), for example through a national or European portal.

This label is a real application of Jonas' *Imperative of Responsibility* (1990), which aims for an ethical judgment on its activities (see Davis and Laas 2014). If it comes to that, it should rather be awarded by an organization independent of the researchers' supervisory bodies, an organization that will have been at the origin of a specific charter applicable to new technologies and for which there is not enough scientific knowledge about the dangers to enter into traditional risk management methods. A number of cases, from nanotechnologies to digital sciences (e.g. AI) could "ideally" serve as demonstrators of this new type of scientific operation. Drawing on the principles of CSR (Wikipedia 2022), the foundations of SRR, elements of sustainable development charters, could be as follows:

– SRR covers social and environmental issues in the work of a research team.

– SRR is not, and should not be, separated from the laboratory's action strategy, since it involves integrating social and environmental concerns into its activities.

– SRR is based on a voluntary concept.

– An important aspect of SRR is the way laboratories interact with their internal and external stakeholders and external stakeholders (employees, customers, immediate environment, supervisors, partners, etc.). A responsible research climate translates into trust and openness to others. It is more a question of dialogical than propagandistic communication (Arendt 1972).

To this end, the exercise of SRR implies a heightened and updated awareness of the environment, as well as respect for balance of interests of stakeholders. SRR respects Koslowski's (1998) reflection: "The ethical fact lies not primarily in suffering the suffering of others, but in paying attention to and avoiding the secondary effects of one's own actions on others". In this sense, the obligatory openness toward what is not directly productive can disturb or even disrupt the dynamic team's research dynamic (like quality in research). What is more, as De Pracontal (2005) reminds us, "the popularizer doesn't say science – it is unspeakable outside its context – he tells stories about science", and possible biases may exist in good faith between what the research says and what the citizen understands or simply perceives. Indeed, scientific jargon is sometimes polysemous and ambivalent, which can lead to misunderstandings.

For their part, Coller and Califf (2009) propose a series of questions for designers that constitute the beginnings of a self-guided global risk assessment tool:

– Does the planned use of the new technology meet a pressing need? What is its added value in relation to what already exists?

– What would be the size and type of market? With what probability of economic viability? How strong is the intellectual property?

– Is there an industrial partner to develop the technology effectively and efficiently? Will the technology have an acceptable price?

– Can a pre-study be designed, with sufficient statistical power or by any other method, to define a robust endpoint (Czajkowski et al. 2016; Pineo et al. 2021)?

Transferring knowledge outside the scientific context is therefore a delicate exercise, requiring time-consuming learning (and not always well recognized by research evaluators). In this context, it must be necessary, in order to change the culture of researchers, we need to send out the signals of a strong will on the part of the supervisory authorities: for example, recognition of an SRR label, training courses, (partial?) modification of evaluation methods support for international accreditation, etc. Should we not reflect on Roman's (1998) words on the ethics of responsibility: "It consists in taking into account the real, the existing, and therefore in renouncing certain convictions that can wait, while not renouncing certainties and long-term perspectives"? The demand for a mutualized approach between management and laboratories assumes that "an altruist can have a selfish prejudice" (Elster 2009), but above all that, the action will develop from a small number of volunteer laboratories who will commit to the process ("Kantian" operating rule) and who will encourage others to imitate them (Laffont 1975; Kreps et al. 1982).

In innovative experimental research, knowledge of the risks associated with materials, chemicals, artifacts, etc., is not always stabilized. This context calls for greater protection of the researcher's working environment, or more broadly of the environment "tout court", to avoid exposing society to potentially unacceptable risks in a context where there is no regulatory framework that are based on risk control. From a factual point of view, decisions based on the use of the precautionary principle can take the form of:

– a ban (temporary or permanent): request for moratorium (e.g. carbon nanotubes);

– a restriction of use;

– targeted information;

– requests for specific in-depth research on risks/dangers for researchers and the general public;

– "simple vigilance";

– a responsible expectation.

In a situation of uncertainty, several criteria should be assessed with regard to the implementation and impact of the planned research program: acceptability, observability and reducibility, on the one hand, and irreversibility, severity and plausibility, on the other hand. These considerations, based as far as possible on all validated knowledge, should reveal the "consistency" of the risk (potentially supported, plausible potential, hypothetical). With this in mind, Chevassus-au-Louis (2007) proposed the use of a hexagonal mapping defined as follows (see Figure 1.6):

– upper part: acceptability;

– upper left: reducibility;

– upper right: observability;

– lower part: severity;

– lower left: irreversibility;

– lower right: plausibility.

Plausibility means an analysis of the scientific "consistency" of the risk, by acceptability the potential attraction of the novelty for the general public; by severity (its potential: population concerned, magnitude, nature of effects, etc.); by irreversibility, the ability of the risk to be temporary; by observability (aspect that can be detected early); and finally, by reducibility, the possibility of taking preventive measures. By assigning a value to each of these considerations, it is possible to situate a "center of gravity" of the possible risk on this hexagon, enabling more informed decisions on the possible risk. For this analysis, leading to appropriate measures, it is possible to implement suitable internal deliberation processes (possibly with external support).

In fact, the consultation of different partners is expressed through a diversity of currents revealing the dimension, complexity and tensions of emerging fields. This can lead to difficulties of "cooperation" between highly heterogeneous players, and potentially the same applies to the application of SRR in communication.

Indeed, for scientists, there are already balances of power between technological, scientific disciplinary purposes, transfer of knowledge to society, environmental and public health, etc. These separate characteristics often rely on an appearance of shared definition. In fact, the new field from a scientific or technological point of view is probably defined at least as much by purposes, "systems of meaning" as by a field of questioning or effective problems or even a list of industrial results.

The existence of a certain vagueness may be due to a lack of clarification on the part of the individual but also to the possibility of exploiting this unstabilized framework to act in a committed manner. In fact, a number of publications have highlighted the gap between scientific knowledge and its limited application in practice (Pfeffer and Sutton 2000, 2006; Rousseau and McCarthy 2007; Rynes et al. 2007; Runu 2021). In such cases, we can expect disfunctions or even breakdowns in dialogue between stakeholders. Observed in cases where uncertainty reigns (e.g. nuclear waste), knowledge of the diversity of representations is a prerequisite for positive and honest exchanges. However, for the most part, when researchers are questioned, they consider themselves as capable, through self-expertise, to act in a way that is adapted to the situation they are exploring (in this context, experience has shown that subjects that fall outside their remit or skills are rather neglected).

In fact, and in agreement with Morel (2002), self-expertise is the result not only of organizations' sluggishness and lack of responsiveness, but also of a culture modestly adapted to new situations involving activities that do not fall strictly within the "hard sciences", activities carried out in a context where the isolation of researchers and the pressure exerted on them suggest total autonomy. Moreover, for many, the difficulty of translating their research into a language adapted to others, a weakness of the will to immerse themselves in activities that are not very profitable and, moreover, at high risk (if certain parameters have been neglected, which risks leading to forms of "tight shielding"). In still infrequent cases, intercultural expertise develops, with classic but manageable difficulties of language, perception, balance of debates (depending on the personal visions of the actors present). This situation is close to interdisciplinary treatments. So, several considerations/centers of interest can then be at work:

– disciplinary considerations: new knowledge, innovation, wealth creation, technical progress, care and medicine, technical risks, toxicity and resource management;

– ethical and moral considerations: prevention, ethics of life, preservation of life, intergenerational solidarity, sustainable development and long-term effects;

– surroundings of daily life: emissions, pollution, hygiene, safety, health, stress, living conditions and comfort;

– framework policy and social framework: priority to employment decisions, civic-mindedness, democracy, neighborhood solidarity (e.g. waste management), "policing" and associated security and terrorism;

– hierarchical considerations (formal or linguistic) leading to forms of silence, even for people who disagree with the proposals of the majority, the one who speaks up. There may be a number of reasons for this, including difficulty in explicitly verifying what has been said, fear of being incompetent, or submission to authority;

– risk of active pooling meetings being transformed into top-down information unacceptable storytelling, avoiding real open debate, a debate which is inherently fraught with uncertainty.

However, based on the best possible knowledge of the diversity of representations (which we should try to quantify (estimate), if only by assigning points to each of the responses to the proposed criteria), there is possibility of "honestly" having a clarification of the role/interests of the various actors with their identity facets, valuing diversity and the possibility of deepening the reflection on particular axes, but also of seeking other partners allowing, as much as possible, to balance the debate.

NOTE.– Integrity/reproducibility.

Plagiarism aside, these two areas are the focus of attention in the scientific world because they lead to non-robust interpretations. Already in 2016, Baker alerted the scientific community to this problem, recently taken up by Diaba-Nuhoho and Amponsah-Offeh (2021). It is no longer a question of being credible and inscribing ourselves in the "right" paradigm, but instead, that of being true. But, as we all know, this means that a certain number of constraints of the New Public Management paradigm (see Figure 1.6).

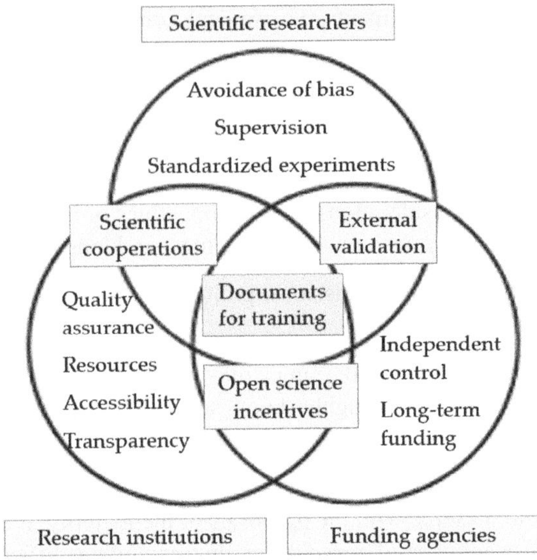

Figure 1.6. *Reproducibility and integrity in research: the role of scientists and their institutions*

In this sharing of a significant portion of ignorance, it must be possible to take stock of the uncertainties of scientific and technical knowledge, approximations of knowledge, misinterpretations and the limits of competence. It must also be possible to measure the extent of unresolved questions and question marks. Within the research team, this approach allows us to distance ourselves from all (?) ideologies, overly reassuring statements and oversimplifications. This means that there is a willingness to attempt to build a real action program from initial disciplinary understandings. The result must be a form of cognitive mutation that will force the rationality of expertise to take account of social demands (or what may take the place of them). This means that players must exploit facts and stabilized knowledge to the maximum, while at the same time taking heuristic account of uncertainty (the problem of building expertise). But to do this, we need a framework that is to move forward.

In this complex approach between scientists and laymen, the neutral aspect does not, in the final analysis, mean a neutral decision, based on the search for a soft consensus. It is possible to experiment to see how the dialogue develops during this period of high cognitive mobilization. Basically, as Reynaud (1989) reminds us, "the heuristic method is a bad method for proving, but a good method for discovering". That is what is at stake in the approach we are proposing, as we try to rediscover meaning and coherence between the different partners involved.

1.3.4. *A few examples*

> All too often people rely on their own (limited) experience to make important decisions, they rely on a few successful examples (often in the sense of benchmarks), and policy makers often rely on one study or only a very few that they happen to have commissioned, and scientists all too often believe that only one or a few good studies really explain everything important about an issue (Frese et al. 2016).

> Incentives, without knowledge and systems, are likely to produce friction and frustration rather than change (Andrew et al. 2021)[4].

> By reversing the perspective on the gap between science and decision-making, we find that local administrators have a different view of this gap to scientists. Administrators seem to have a more nuanced, if not completely opposite, perception of the different epistemic contexts of scientists and decision-makers, of the uncertainty inherent in scientific knowledge and the rationality of decision-making (Van Stigt et al. 2015)[5].

To go further in practical terms, the precautionary principle is associated with the French "Barnier" law 02-02-1995 – articles L.200-1 of the rural code and L.110-1 of the environmental code – summarized below:

> The precautionary principle, according to which the absence of certainty, given *current* scientific and technical knowledge, must not delay the adoption of effective and proportionate measures aimed at preventing a risk of serious and irreversible damage to the environment at an economically bearable cost.

In innovative experimental research, knowledge of the risks associated with materials, chemicals, artifacts, etc., has not yet stabilized. This context implies reinforced protection of the researcher's working environment, or more broadly of the "just" environment, to avoid subjecting society to potentially unacceptable risks in a context where there are no regulations in place based on risk control.

From a factual point of view, decisions based on the precautionary principle can result in:

– a temporary (or definitive): request for a moratorium (e.g. carbon nanotubes);

– a restriction of use;

4 Author's translation from the French version of the text.

5 Author's translation from the French version of the text.

– targeted information;

– specific in-depth research on risks/hazards for researchers and the general population;

– "simple" vigilance;

– irresponsible waiting.

In a situation of uncertainty, several criteria should be assessed with regard to the implementation and impact of the planned research program: acceptability, observability, reducibility, irreversibility, severity and plausibility. These considerations, based on all validated knowledge, must reveal the "consistency" of the risk (potentially substantiated, plausible potential, hypothetical). Based on these considerations, Chevassus-au-Louis (2007) proposes the use of a hexagonal map as shown in Figure 1.7.

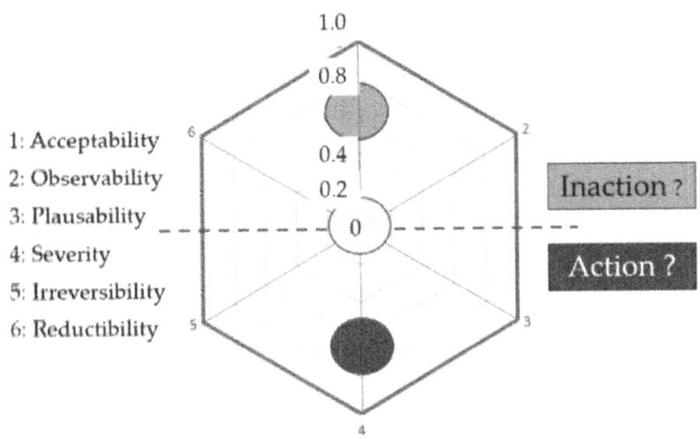

Figure 1.7. *Risk mapping for a decision in a situation of uncertainty*

Plausibility means an analysis of the scientific "consistency" of the risk (see above); by acceptability, the potential attraction of the novelty for the general public; by severity (of the risk) its potential: population concerned, magnitude, nature of effects, etc.; by irreversibility, the ability of the risk to be temporary; by observability (early detectability); and finally, by reducibility the possibility of taking preventive measures. By assigning a value to each of these considerations, it is thus possible to situate a "center of gravity" of the possible risk on this hexagon, enabling more informed decisions on the possible risk (but without quantifying it). For this analysis, leading to appropriate measures, it is possible to implement suitable internal deliberation processes (possibly with external support). On the basis of these exchanges, the unit/group may consider carrying out or requesting further research, reinforcing its scientific watch, or, more prosaically, considering stricter safety instructions, calling in external advice, etc. Two examples, taken from de Bouet (2009), illustrate the possibility of using this proposal (see Figures 1.8 and 1.9).

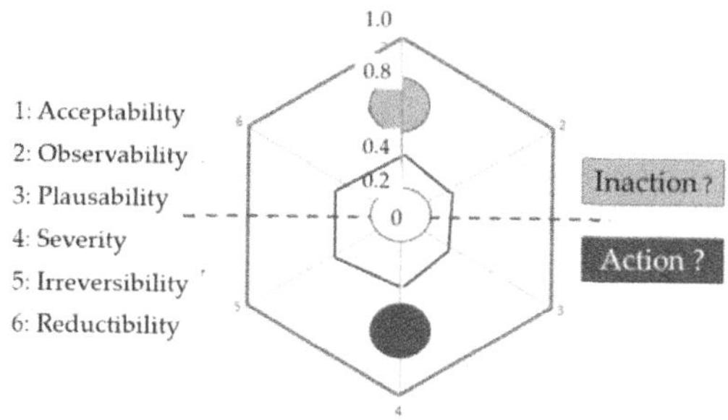

Figure 1.8. *The case of mobile telephony*

In the case of Figure 1.8, produced in 2009, this mapping does not lead to a clear-cut opinion; on the other hand, in the case of the handling of the "mad cow" crisis, the center of gravity of the polygon of risk indicators has shifted sharply downwards. At the time, this led to a radical solution involving entire herds, particularly in the UK.

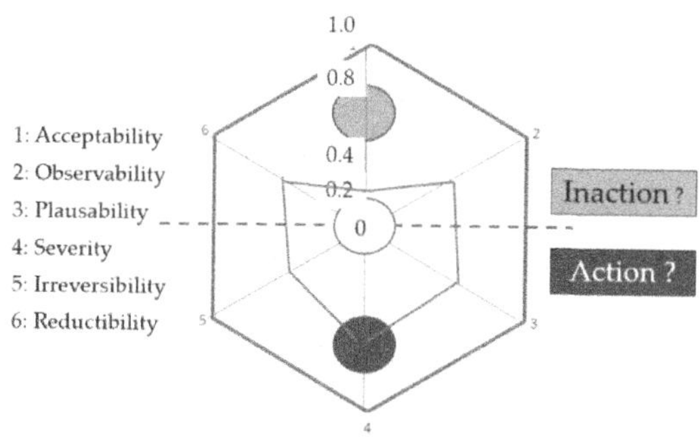

Figure 1.9. *Case of bovine spongiform encephalopathy*

Concerning the use of nanoparticles, in 2009 (HCSP 2009), the health watch group on the health impacts of nanotechnologies, attached to the school of the HCSP (*Haut Conseil de la santé publique*, French High Council of Public Health), recommended, on the basis of the precautionary principle, that the production of carbon nanotubes and their use for the manufacture of intermediate products or consumer products and health products are carried out under strict confinement conditions. It also proposes actions for identifying potential exposure and nano-safety situations, in particular for carbon nanotubes, and avenues for research on metrology and risks. In this same High Council, it was possible to work on the risks induced by other nanoparticles such as nano-silver.

The HCSP (2010) has also examined the question of the safety of silver nanoparticles widely proposed for the elimination of microbes, bacteria and/or viruses. The opinion draws the attention of health authorities to a series of measures that should be carefully followed. These recommendations relate, in particular, to the monitoring system, product labeling and research that should be pursued in greater depth.

Finally, given that titanium dioxide nanoparticles are widely used in food, cosmetics and construction materials, and that they are manufactured or handled at numerous sites in France, which raises the question of exposure of workers and local residents, the HCSP (2018) has issued a recommendation to the French Ministry of Health. TiO_2 particles can contaminate the atmosphere, requiring specific protective measures, because these nanoparticles, inhaled or absorbed, cross biological barriers and accumulate in certain organs. However, the health risks are still uncertain. The HCSP therefore recommends:

– the protection and medical monitoring of workers handling these nanomaterials, in particular pregnant women and women of childbearing age;

– the application of good preventive practices in the workplace with regard to the production, handling, storage, packaging, transport and integration of TiO_2 particles into finished products, as well as recycling;

– protection of local residents and the environment around sites where TiO_2 nanoparticles are produced, stored and handled;

– the development of metrology and exposure monitoring in the workplace and the environment;

– the adoption of specific exposure limit values for TiO_2 nanoparticles;

– reinforcement of the obligation to declare the presence of nanoparticles in products and materials in the R-Nano database, by including substances with a minimum nanoparticle content of 10%, and making this database accessible to all organizations required to assess the risks of nanomaterials;

– that manufacturers and importers of TiO_2 include in their dossiers submitted for expert appraisal under REACH system studies on the reprotoxic potential of TiO_2 nanoparticles;

– support for research into the mutagenesis, carcinogenesis and reprotoxicity of TiO_2 nanoparticles;

– strengthening the resources of prevention organizations (INRS, Santé publique France, INERIS, ANSES) in terms of expertise on nanoparticles.

NOTE.– Acronyms.

– REACH: EU regulation adopted to better protect human health and the environment from the risks associated with chemical substances, while promoting the competitiveness of the EU chemical industry.

– R-Nano: enables manufacturers, importers and distributors of more than 100 g of substances in nanoparticulate form (nanoparticles, nanomaterials or nanotechnologies) to make mandatory declarations concerning the quantities produced, imported or distributed (see ANSES activities).

– INRS: *Institut national de recherche et de sécurité* (French National Institute of Research and Security).

– *Santé publique France*: national public health agency, a French public administrative body under the supervision of the Ministry of Health.

– INERIS: *Institut national de l'environnement industriel et des risques* (National Institute of the Industrial Environment and Risks).

– ANSES: *Agence nationale de sécurité sanitaire de l'alimentation, de l'environnement et du travail* (National Agency for Food, Environmental and Occupational Health Safety).

Figure 1.10, which uses this mapping for these different nanoparticles, carried out with a group of volunteer students who were informed of the results of the literature from ENSIC (*École d'ingénieurs en génie des procédés* of the University of Lorraine), enables us to examine the differences in terms of the risks associated with these materials. The results are in line with the HCSP's three recommendations.

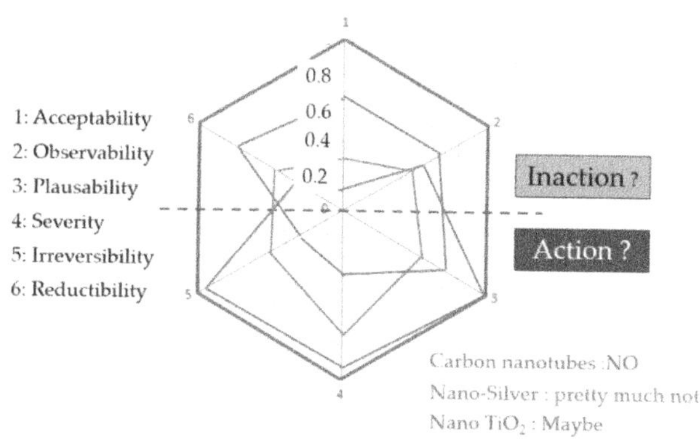

Figure 1.10. *Example of nanoparticles*

A group of experts must express their views on the interconnection of systems and unstabilized issues, taking into account as far as possible the effects and feedback on (current) practices of thought, science and action and so on. Expertise may find itself powerfully incited to undergo the anxiety generated by the emerging problem, forcing it to think urgently. With the Covid-19 crisis, a model, with unstabilized data, even absent from the pandemic, was carried out (Reversat and André 2022). Previous students were able to reflect on mapping applied to different citizen protection situations. The results are shown in Figure 1.11. However, according to Theys and Cornu (2019), several categories of questions seem to require priority attention:

– how scientific knowledge can be combined with forecasting and foresight;

– the choice of time "horizons", the determination of critical thresholds for action, and the taking into account by both scientists and stakeholders of the "time remaining to act";

– the concepts of "adaptation" to the perceived situation, the different meanings, the role of politics, international representation, etc;

– evaluation and historical analysis of the successes and failures of past attempts to frame collective action in time;

– reflection on decision-support tools, such as the prioritization of major irreversibilities or the discount rate.

These areas, as expressed in Figure 1.11, were only partially discussed with the students, for lack of time. In addition, the Covid-19 crisis presents society with the plan for the future in a context of extreme uncertainty. And yet...

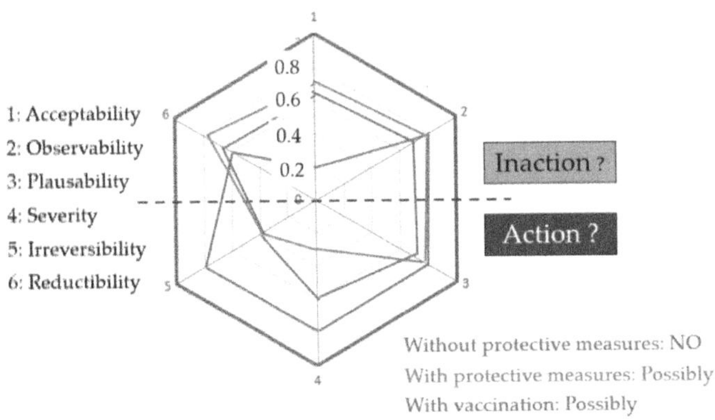

Figure 1.11. *Example of the Covid-19 pandemic*

NOTE.– Expertise.

In fact, the consultation of different partners is expressed through a diversity of currents revealing the dimension, the complexity and the tensions of the emerging fields. This results in difficulties of "cooperation" between highly heterogeneous actors. The same must apply to the application of SRR in communication. Indeed, for scientists, there are already power relations between technological, disciplinary scientific, transfer of knowledge to society, environmental and occupational or public health purposes, etc. These separate characteristics often rely on an appearance of shared definition. In fact, the new field from a scientific or technological point of view is probably defined at least as much by purposes, "systems of meaning" as by a field of questions or actual problems or a list of industrial results, etc. In these cases, we can expect malfunctions, or even breakdowns in the dialogue between stakeholders. As a result, in cases where uncertainty reigns (nuclear waste, GMOs, augmented human beings, for example), knowledge of the diversity of representations is a necessary prerequisite for the action of positive and honest exchange. Several considerations/centers of interest, repeating some of those expressed above, may be at work.

– disciplinary considerations: new knowledge, innovation, creation, technical progress, care and medicine, technical risks, toxicity and resource management;

– ethical and moral considerations: prevention, life ethics, preservation of life, intergenerational solidarity, sustainable development and long-term effects;

– framework of daily life: emissions, pollution, hygiene, safety, health, stress, living conditions and comfort;

– framework policy and social framework: priority to employment decisions, civic-mindedness, democracy, neighborhood solidarity (e.g. waste management), "policing" and associated security and terrorism.

Responsible research and thus refer to ways of proceeding at an early stage, obtaining relevant knowledge about the (possible) consequences of the outcomes of their actions and about possible scenarios, evaluating foreseeable outcomes and options in terms of moral values and using these considerations as functional requirements for the design and development of new research, products and services (Davis and Laas 2014).

Based on the best possible knowledge on the diversity of representations (which we should try to quantify, if only by assigning points to each of the responses to the proposed criteria), there is the possibility of clarifying role/interests of the various players, to value diversity and the possibility of deepening reflection on particular axes, but also to seek out other partners to balance the debate as far as possible. In this sharing of a significant proportion of ignorance, it must be possible to take stock of the uncertainties of scientific and technical knowledge, the approximations of knowledge, the abuses of interpretation and the limits of competence. It should also be possible to measure the extent of unresolved questions and question marks. This approach should enable us to distance ourselves from all (?) ideologies, overly reassuring statements and oversimplifications.

To do this, you need time and a sufficiently composed or partially "neutral" framework in which to move forward. But neutrality does not mean neutrality in decision-making based on the search for a soft consensus.

1.3.5. *Authoritarian constraint*

> Man is not born carrying within himself the meaning of his life; we must necessarily conclude that no human society is possible where the normative arrangements have not been imagined and instituted, thanks to which the subjects of successive generations attain the status of human beings (Michéa 2006).
>
> Many moral philosophers and other thinkers live in a peaceful world, barely obscured by a few bands of cloud, and as they walk along the sidewalks that take them from their offices to their homes, they dream deliciously of intentionality (Richards 1993)[6].
>
> [...] it is unlikely to be useful if the policy makers get too involved in the decisions on how these studies should be done, but they should rather encourage competing methodological approaches of evidence development to allow for triangulation (Frese et al. 2012).

In 2008, the European Union published a document entitled "Commission Recommendation on a Code of Conduct for Responsible Nanosciences and Nanotechnologies Research". This document, which is intended to be general, attempts to translate into constraints, a number of principles to which the entire research community subscribes (even if they are only recommendations). In this new context, the need to change culture is perfectly well felt, but cannot be produced in such an "angelic" (?) and ultimately radical way as that presented by the Union in the document. Initial training, the disciplinary scientific approach, staff evaluation methods have shaped minds in all (or virtually all) areas of "academic" research. If exploring the social field is a priority, the approach needs to be validated, assimilated and appropriated. In short, it will take time (and money) for it to become an integral part of researchers' (good) practices. Indeed, it will be necessary to

6 Author's translation from the French version of the text.

disrupt the cognitive processes that will enable the culture of both researchers and stakeholders to evolve, leading to the adoption of more socially responsible behaviors (by accepting stops, revisits of scientific activities, questioning, etc.).

Moreover, in the field of exploring the dangers to health and the environment induced by nano-objects, France (as well as Europe) has a very (or too) limited number of good toxicologists capable of addressing these essential themes (remember that when the REACH procedure was set up, only a few percentages of the dangers of pure chemical substances used industrially were well known on a world scale, whereas the development of chemical dates back to the second half of the 19th century!). Under these conditions, where the need for technical progress is desired, research teams from various R and R&D organizations (CNRS, INSERM, IRSN, INRS, CEA, universities, etc.) and industries have shown that it is possible to define effective collective and individual protection of the various operators of research laboratories. In the absence of knowledge of the dangers, it is, in a responsible manner, preventive operations that must therefore be favored. In principle, there is already a way to avoid major problems at the design stage (up to the POC).

In a code of conduct designed to establish a calm debate on “nanoparticles”, which is intended to serve the “well-being” of society as a whole, it may be advantageous to present possible risks, uncertainties, possible irreversibilities and the potential benefits of their development in a balanced way. Now, if we were to demonstrate the harmlessness of electromagnetic waves from cell phones, the advantages of this freedom (?) of connecting with other citizens be destroyed/forbidden in the name of a low health risk? The vague concept of well-being introduced in the European document is insufficient to define appropriate rules. It is all about acceptability. We are well aware of its shortcomings, in particular its potential for manipulation, its temporal evolution and its sensitivity to “social correctness”.

We can clearly see this today in the war between Russia and Ukraine, which is leading to sudden cuts in the supply of gas to Europe with a reversal of opinion regarding nuclear power. This simple situation dulls the entire operational scope of this European code. Moreover, in the field concerned, speaking generally of nanoparticles is like speaking of chemical risk, not all chemical substances are dangerous in the same way; globalizing is not solving the problem. It is therefore at the stage of the research team or of a group engaged in the definition of a pre-POC that it is necessary to integrate all of the problems associated with specific innovations, but not only for nanotechnologies alone, real point of crystallization of social concerns around 2010. These, as illustrated by André (2017) and Schaer et al. (2020), have largely evolved due to the lack of reliable data in sufficient quantities, the weariness of opponents and the temporary existence of other possible risks. One risk drives out the other, without always being treated.

The first point of the document states that this code "invites all stakeholders to act responsibly...". Clearly, it is not possible to develop "Mode 2" research in the sense of Nowotny et al. (2003, 2005), who hypothesize a concerted, "harmonious" choice between science, technology and society to develop "consensual" research. Various surveys carried out in the United States and Europe show that the public is quite unable to keep up with scientific developments, and probably even biased. When it comes to risk, are researchers able to plan ahead? Have they been trained to express risks heuristically?

Thus, and this is the whole SRR debate, feedback is due not only to the stakeholders, but also to the public, whose (difficult) understanding of the uncertainties linked to innovations should enable "peaceful" monitoring, rather than crises supported by certain players exploiting (too) many uncertainties. Unfortunately, not everything can be fully understood by the public. Translation is tricky and can leave room for many interpretations. This can lead to blockages and crises, as the irrational may be called into play (another risk management to operate). Nisbet (2009a, 2009b) considers it important in order to move beyond "polarized" and compartmentalized interpretations of an open question to engage voluntarily in dialogue with society. This means remaining faithful to scientific knowledge, but also avoiding denigration of the social groups involved. In any case, this dialogue cannot be seen as a strategic attempt to "sell" science (reverse storytelling), which is practiced even by consulting firms paid to develop language adapted to "requalifying" debates (Nisbet and Scheufele 2009); the aim is indeed to recognize the points of agreement between stakeholders and to understand each other on those at the root of disagreements, thus authorizing a process of progress.

The European document seems marked by a concern for fear of being accused of not having identified all of the potential problems linked to the application of "nano" research, leading to the assumption today of irresponsible activities on the part of the research and even industry. Beck (2003) and Jonas (1990) reminded us that we live in a society with risks and that the sacrosanct zero risk does not exist! Responsibility consists of making choices and assuming them at all levels, and it is indeed the goal of a democracy to commit to meeting the needs of society (avoiding as much as possible, naturally and obviously and reaffirmed, major and irreversible effects).

Moreover, as far as research is concerned, should we not rather speak of a precautionary attitude in the recommendation rather than the principle which comes under interdisciplinary expertise, carried out by scientists who reason in "complete non-knowledge of cause", in general, under time pressure and, then, validated (or not) by the state? We can, at the limit, understand that, on a vast social subject (apart from research) such as that of "mad cow disease", it was necessary to have a decision-making aid, because the problem was important (but scientifically localized) and likely to lead to a serious crisis; but how will we manage to define it on a European scale, for each small specific case in laboratory research, the (good) rule to follow… It is indeed at the stage of the research unit or the pre-POC that it is necessary to act and to change the culture of the personnel. Thus, on this proposal, it is possible to make the following comments:

– There is no advantage in defining a code solely for the "nanos" component; information science and technology, information technologies, organizational changes, etc., have changed behavior to a very large extent without anyone being concerned, so why nanos? Because the invisible risk is more worrying than the known risk? To avoid blocking research, as is the case with GMOs?

– We must take into account the timeframes and wishes of decision-makers to achieve effectively responsible research with the costs that go "with it"; a cultural change cannot be decreed, it must be built.

– The European Union must set an example by providing significant funding to achieve the proposed objectives.

– The document says too much about what we must or must not do, and fails to explain how we are going to achieve the humanist objectives (unchallenged, because they cannot be challenged) recalled in this chapter.

– Do we not already have the necessary arsenal from an ethical, legal, regulatory arsenal that makes this anxiety-inducing, paranoid text useless for all parties?

So, in an attempt to be unassailable, reading this code gives the impression that nanomaterials only entail risks for mankind, whereas there are benefits already being industrialized and others, as yet unimagined, that could emerge. Naturally, there is a need for increased and shared vigilance, but it probably does not require all of this formal and impractical paraphernalia. Besides, by 2022, we will not be hearing any more about this "umbrella" document that only protects those who wrote it. It is another form of responsibility! Moreover, scientific research must not suffer the effects of a proposal from the Union that imposes obligations of results without the responsibility being shared at least with the European principals, some of whom contribute to research funding.

In this sense, this type of document is ill-suited to sharing social issues as part of a dynamic of innovation that could bring jobs, purchasing power, quality of life and a sense of community in a Europe that is still searching (even if changes are taking place with the Russia–Ukraine war). It is all the more dangerous because, after a quick read, the positive vision conveyed by the document (but which exists in other equally legitimate frameworks) will probably be retained by a reader in a hurry relative to its inoperability. If extended consultations are carried out under time pressure, it is reasonable to assume that responses to these requests for opinions will be positive, with very few remarks. The "soft" majority will prevail over the minority who have thought things through because they are responsible.

The construction of a shared collective imaginary value-added of good governance must be one of the concerns of the European Union, which, from the author's point of view, probably has other things to do than to take refuge in a questionable appraisal of the future of nano-electronics, which is either ill-adapted to reality or biased, in order to satisfy who knows what group of players? Or, as mentioned above, to avoid being accused? The latter hesitates between an understanding of the document that respects a form of angelism associated with a "good conscience" (with, however, a transfer of responsibility!) and a possible more cynical intention to achieve/maintain a fake social peace. Where then is the added value and responsibility of the authors of this document?

The author's approach, as we have come to understand, is aimed at a commitment to the voluntary nature of certain responsibilities, forcing those involved in (necessarily risky) research to justify and explain their choices. The second, that of the document, consists more in denouncing a delegation of responsibility to these actors (Charbonneau and Estèbe 2001). According to Appel (2001), we find ourselves in a situation of "bifurcation architectonics" between moral standards and legal principles. In the former case, the aim is to promote the capacity for responsible autonomy, linked to ongoing evaluation by the researchers themselves, and more likely to be evaluated periodically and/or a posteriori by those in power.

In this situation, we come across the old debate between freedom and autonomy on the one hand, and obligation and duty on the other. Perhaps politics also does not want to be relieved of the fate of society by scientists (Marinese 2008), which is laudable in itself, even if the means to achieve the objectives are not defined. "Doesn't this rationalization of lawmaking [recommendation] ignore the irrationality of facts?" (Bergel 1993).

In this situation, we find an old debate between freedom and autonomy, on the one hand, and obligation and duty, on the other hand. The choice of autonomy is linked to a need to trust, because no "regulator" has day-to-day knowledge of what is done in the laboratory or in the informal group in charge of the pre-POC, nor of what is being prepared in the head of the researcher or the creative who has not yet been forbidden to think. Common ground between the two visions may exist, but it must be wanted in the name of a negotiated benefit for humanity. Moreover, in trying to move in this direction, it is useful to remember that our society, with all of its complex and evolving interdependencies, is defined less and less in terms of roles and statuses. It is then difficult to assign an obligation to a single "target". It therefore seems important to enter into joint processes of sharing responsibility in a true spirit of consultation and cooperation. The Union document should, in this sense, commit to the consensual establishment of meaningful partnership arrangements.

1.4. Provisional conclusion

> Inventors tend to have their most impactful innovations around age 40 and their incomes rise rapidly just before they have high-impact patents (Bell et al. 2019).

> Decision-making, which can influence the structure of the social network is the need to create a consensus. As Whyte et al. (1989) note, the first task of all decision-making groups is 'to produce a consensus from the initial preferences of its members' (Kijkuit and van den Ende 2007)[7].

> The ideas of economists and political philosophers, both when they are right and when they are wrong, are more powerful than is commonly understood. Indeed the world is ruled by little else. Practical men, who believe themselves to be quite exempt from any intellectual influences, are usually the slaves of some defunct economist [...] It is ideas, not vested interests, which are dangerous for good or evil (Keynes 1953).

Beck (2003) reminds us that "it is no longer the scale of the risk that is changing, but its 'scientification', which no longer allows us to offload our responsibilities onto nature". Researchers can no longer remain in their cozy disciplinary silos and must, like everyone else, take an interest in the world by returning, albeit modestly, to a less modular, less mono-disciplinary, less mutilated production. To do so, it must participate in the coordination of production, with a view to achieving operational efficiency of interest to society (Berger 2006) and, consequently, open up to a culture shaped by the reductionism of linear thinking (Mariotti 2008).

It is not a question of defining a stable path for action, but rather a commitment to openness. Indeed, not everything can come from this one, ultimately privileged world; we must also bear in mind Girard's (2002) very important remark on society. He writes: "It's not that I believe in an intrinsic superiority of the still new fields of our culture – science, technology, economics. But, in my opinion, our cultural

7 Author's translation from the French version of the text.

activities are vulnerable in proportion to the spiritual greatness that should be theirs". SRR is probably one of the ways in which we can get closer to an anxious, ill-educated, ill-informed society with changing desires. It is, moreover, on the basis of this rapprochement that it may be possible to avoid untimely and random controls and crises, but this requires a responsible attitude that is ultimately new in its expression.

In any case, this radical change requires a real social will if we are to make research a socially useful tool for the long term, in a world that is exploring new paradigms. Yet, as Einstein (1991) wrote, "no problem can be solved without changing the mind that created it". This strong agreement runs the risk of being a rhetorical device using the lazy, easy route to "sloganized" imprecations that essentially boil down to a display. As we have said, it's as if it had already been done! In fact, this is one of the criticisms levelled at national politicians. All we need to convince ourselves of this is the proclamation of a desire, not really operational, to reduce the production of carbon dioxide for the environment. In a particularly important case, the European Union published a "nanocode" of good conduct on nanotechnologies (2008), highlighting a genuine desire to avoid any risks to people and the environment, which could lead to discernible blockages. Is public pressure (the possible asbestos effect) leading the EU to engage in pseudo "political correctness"?

In the absence of a real will on the part of all the stakeholders, which is quite likely, it will unfortunately always be possible to carry out fortunately useful disciplinary work. But how, and on what ethical foundations, will we define the future conditions of social utility enabled by the results of research in terms of invention?

According to Peyrat (2003),

> Socialization now requires a real project, to be deliberated, implemented and explained: the institution of a new urbanity. On a societal scale, we can no longer rely on automatisms to guarantee mutual trust. We need to create new conditions for its emergence.

Does the causal and enlightening case of new technologies, supported by all the funding bodies, basically not risk overshadowing or, on the contrary, revealing this necessary revisiting of the role of research into short-, medium- and long-term risks, in a world that is increasingly escaping the complex and ambivalent desires of mankind?

> Our entire destiny, with all its facets, depends on this ability [to meet challenges]. The reason for this is simple: can we for a single moment expect our culture, the values we hold dear, our potential for influence, our art of living, would not gradually fade away, if we were no longer able to take our place among the living, to take part in the

> global struggle, in the scientific, economic and technical fields? (Leprince-Ringuet 1980).

Another vision of responsibility...

These various elements, which are explored in depth in this chapter, need to be set against the processes that lead to invention and innovation. For example, as expressed by Aghion et al. (2020), there is pressure for growth, with renewals that can be radical (Schumpeter's creative destruction). Innovation is based on a cumulative process (waves of secondary innovations) in a favorable environment that allows competition. Even if we can envision a trusting dialogue between innovation and SRR, the result of this expertise, which we hope to share, can lead to failure (as in the case of the industrial development of carbon nano-tubes in France, for example). In this sense, the SRR axis is a "death valley"!

Death Valley No. 12: passing SRR's "filter" favorably

> Responsible research and innovation refers to ways of proceeding in research and innovation that enable those who initiate and are involved in research and innovation processes at an early stage (A) to obtain relevant knowledge about the consequences of the outcomes of their actions and about the range of options open to them and (B) to evaluate outcomes and options effectively in terms of moral values (including, but not limited to, well-being, justice, equality, privacy, autonomy, safety, security, sustainability, responsibility, democracy and efficiency) and (C) to use these considerations (under A and B) as functional requirements for the design and development of new research, new products and new services. (DGRI 2013)

In this context, is Leblanc (2020) right or wrong when he writes: "we have no choice but to keep fighting to be among the winners. But beware of the pacifists of the economy [supporters of SRR]: like all pacifists, they are just good at losing wars, not at winning peace!"?

1.5. Conclusion: from the 12 "Labors of Hercules" to the 12 "death valleys"

> We're dealing with the ideological whenever a signifying production is considered in relation to the basic mechanisms of social functioning as constraints on the generation of meaning (Veron 1979).

The creative imagination:

– "disappointing part of Man [...], mistress of errors and falsity [...], superb power, enemy of reason, which likes to control and dominate it" (Pascal);

– "error and disorder entering the mind, at the same time as the tumult of the body" (Alain);

– "essential poverty because the passive object is kept alive artificially" (Sartre; Gust-Desprairies 2003).

> I told you that [Man's] freedom consists of his power to act, and not in the chimerical power of wanting to want (Voltaire 2009)[8].

Hercules is the son of a god (and not just any god, he is Zeus) and a mortal (but still a princess), so he's a demigod. The gods of Olympus are complicated characters who intervene in all the world's affairs, making Hercules irresponsible but guilty of the death of his wife and children. He is therefore condemned to perform 12 impossible tasks as a punishment! He has the intelligence, courage and strength to fulfill his destiny. As with Hercules, not all the information needed for the path ahead (in this case, innovation) is available, and his success depends on mastering uncertainty and risk. That is where my recollection ends, but it is close to the 12 "valleys of death" evoked in Chapters 1 and 2 of Volume 1 and Chapter 1 of Volume 2, and recalled in Table 1.2 (see also Berkun 2007). These are not all consensual proposals, because some components of society (from research to the public) may (have) different cultures, desires and political visions, not all of which can be easily reconciled either between compartments, or within at least one of them. The polydispersity of points of view is one of the characteristics which means that an idea can become profitable or be rejected.

NOTE.– Interdisciplinarity.

"We can define 'interdisciplinarity' as that social phenomenon in which different epistemic cultures (Knorr-Cetina 1999) meet and dialogue, exchanging tools, theoretical and normative frameworks, as well as know-how, and often sharing common practical goals" (Di Vigni 2021).

These various interdependent items are shown in Figure 1.12, adopted from Kammoun and Le Bas (2020).

---

8 Author's translation from the French version of the text.

"Death Valleys"	Comments
No. 1: having a "good idea"	"Brilliant!" I've got an idea (am I a demigod?), and at this stage, is it recognized as such (by the sender, or even by others)? Having a good reason for a creative idea. Providing a framework to generate ideas (they can come from anywhere, in any work environment). Make people "dream" with divergent thinking.
No. 2: being able to integrate a convergence/divergence system	Going beyond the simple idea; giving it coherence and envisaging an inventive project (thought in terms of systemic processes); complexity. Considering known partners.
No. 3: an environment that supports convergence	Talking about it to give substance to the project (an idea is one thing, a functional solution is another), while maintaining a minimum of confidentiality, and involving known partners on the basis of availability, openness, scientific and/or technological relevance, etc. Joint approaches between the cultures of philosophy and engineering cultures.
No. 4: (still) having a creative idea, despite formal training...	Not being shy. See No. 1: between "real" training and the possibility of wandering, if possible "learned", while remaining credible or being "taken seriously". Risk-taking between disruptions not easily financed (associated risks) and traditional financial support for incremental activities.
No. 5: having a creative idea that is recognized as such **N° 5 bis: having a creative idea and integrating it into a rational approach**	Idea desirability. Validly associate partners to carry out a pre-proof of concept (specifications and heuristic approach). Paradigm and innovation process as proposed by G. Dosi as early as 1982. Invention from elements in complex interactions and systems analysis in an interdisciplinary context; complexity sharing and pooling through encounters, dialogues and exchanges of tools, theoretical and normative frameworks; integration. Thinking short-term for companies. Knowledge of the real and the possible. Breaking down and/or bypassing barriers. Considering a forward-looking approach.
No. 6: creating a POC in an interdisciplinary framework	To be financially supported and have the people and resources to carry out a POC heuristic approach and complexity. Designing a technology with markets that correspond to a "winning lottery". Finding the right path in a sprawling administration, the administrative maze is still in place, and even growing ("In 1980, Jacques Dominati, Secretary of State for the Civil Service, felt that we French had the best administration in the world. The proof: foreign delegations were constantly coming to our country to learn from us. Forty years later, his distant successor couldn't make the same remarks without triggering general hilarity, as the many recent examples speak for themselves" (D'Haraucourt 2021)). Patent application.

No. 7: supporting to reach the POC and beyond	Maintaining desirability; deciding to continue or stop. Major transition in complexity management to achieve the simplest proposition possible. See No. 6, but with the option of creating an industrial demonstrator (ID) taking into account socioeconomic criteria with feedback. KIS; ease of implementation; compatibility.
No. 8: existing and being supported globally by society	An approach recognized by the business world and the means to achieve it; potential desirability of the invention. See No. 6; the question of the market is raised. Risk-taking to create a market segment that did not exist before the new technology associated costs. Needing to reach the people for whom the innovation is perceived as a benefit; subsidiarity ease of implementation; compatibility. Estimating the effects on society performance, employment, environment, etc.
No. 9: access to co-creation	Decision-making. Finding partners (noncultural technicians) to help transform an invention into innovation moving away from a "linear" model to one in which two types of players are both co-producers and co-users of information and knowledge. Using heuristics to bring about the best change in a poorly understood situation and within the available resources. Staying on course with the allocated budget. What type of training is required before industrialization?
No. 10: arriving on time	Decision-making. Overcoming our obsession with the past. Numerous filters slow down the process, which can mean that an innovation to reach the market too late; the question of standards is part of this valley. Possibility of trial runs prior to production launch. What training is required before industrialization. The agreement concerns an ID with its entire application environment completed. The innovation was perhaps too revolutionary for its time. Knowledge of opposing processes and the implementation of their ideas.
No. 11: reaching the market	Effective desirability of innovation decision. Jungle of public support for innovation. Resolving the issues already raised. Possibility of pre-production testing and observability (are the effects of innovation visible?). Building a business model. Creation of a start-up, integration into a company's industrial production, etc. (outsourcing of R&D, etc.).
No. 12: passing SRR's "filter" favorably	Ethical activity to be taken into consideration; effects of innovation on society. Satisfying regulations (France is said to be the country with 400,000 standards!). Environment.

Table 1.2. *The 12 "death valleys" and associated criteria*

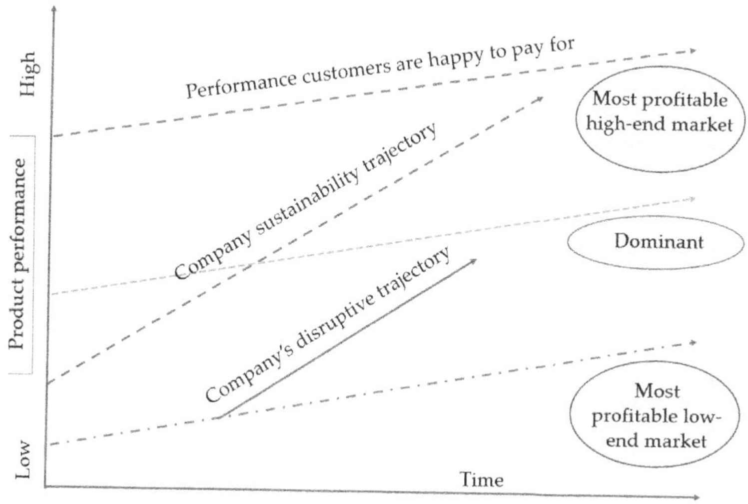

Figure 1.12. *Innovation trajectories*

Following the example of microelectronics with its Moore's Law (which is starting to lose its validity), the Western world succeeded in improving its technologies during and after the Industrial Revolution. In recent decades, the historiography of technologies has focused on radical breaks with the recent past. Within this temporal framework, the academic world has, according to Edgerton (2010), moved beyond a "thoughtless discourse on progress to become (in theory and in theory) eclectic, playful with time and open to marginality". He completes with these "definitive" sentences:

> We should be wary of favoring the novel, both in history and historiography, not out of conservatism, but because we understand the power and influence of claims to novelty to mask a lack of novelty, and sometimes even to suppress it. The promotion of novelty is itself profoundly lacking in originality, but old-fashioned futurism, though outdated, remains surprisingly effective. As [...] the Comte de Lautréamont said long ago, progress implies plagiarism but, we might add, not all plagiarism is progressive[9].

For Corichi García et al. (2018), no innovation is intrinsically disruptive at the outset; it is only through companies making strategic choices to position an innovation in a disruptive way that development takes place. While the first part of this two-volume work (Chapters 1 and 2 of Volume 1 and Chapter 1 of Volume 2)

9 Author's translation from the French version of the text.

showed the existence of many important factors associated with innovation, on a case-by-case basis, we find great differences between these criteria, so that in agreement with Taalbi (2017), a universal theory of innovation seems unrealistic.

As "keys to reading the world, the quality of goods and their accessibility constitute a privileged field of observation for reflecting on the future evolution of our societies". This predictive tendency is as much about "projecting contemporary concerns, hopes and events into the future" as it is about "acting on the present" (Belasco 2006). This is what we examine when we look at additive manufacturing technologies, presented below in the second part of this two-volume work (Chapters 2 and 3 of this volume).

NOTE.– In the South, little salvation?

Van de Vliert and Murray (2018) have produced an analysis of creativity that takes into account the intellectualization, industrialization and urbanization of society. Their results, shown in Figure 1.13, indicate a correlation that higher thermal demands hinder creativity in poorer populations, but promote it in wealthier ones.

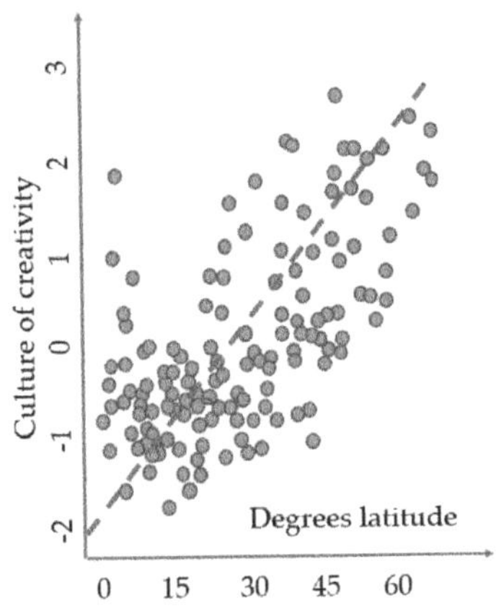

Figure 1.13. *Correlation between latitude and creativity*

1.6. References

ACIA (2008). La communication des risques et le gouvernement [Online]. Available at: http://www.inspection.gc.ca/scripts/priimp/priimppage.php?lang=fsci=www.inspect.

Aghion, P., Antonin, C., Bunel, S. (2020). *Le pouvoir de la destruction créatrice*. Odile Jacob, Paris.

Agora (2008). Principe de surveillance des nanotechnologies et nanomatériaux [Online]. Available at: http://fsm-sciences.org:80/spip.php?article184.

Allum, N., Sturgis, P., Tabourazi, D., Bruton-Smith, I. (2008). Science knowledge and attitudes across cultures: A meta-analysis. *Public Understanding of Science*, 17, 35–54.

Andersen, I.E. and Jaeger, B. (1999). Danish participatory models: Scenario workshops and consensus conferences: Towards more democratic decision making. *Science and Public Policy*, 26, 331–340.

André, J.C. (2000). What does prevention research have to offer the working environment? *Arbëte Och Hälsa (Sweden)*, 18–31.

André, J.C. (2005). Complexity and occupational safety and health prevention research. *Theoretical Issues in Ergonomic Science*, 6, 483–507.

André, J.C. (2006). Weakness of scientific and technical culture of the public and consequences for the development of prevention actions. *Theoretical Issues in Ergonomic Science*, 7, 447–468.

André, J.C. (2008). Vers le développement d'une recherche durable… Ou vers la (ré)humanisation des sciences des artéfacts. *Environnement, risques et santé*, 7, 47–54.

André, J.C. (2013). Towards a socially responsible research (SRR) charter in engineering sciences at CNRS level. *International Journal of Technoethics*, 4, 39–51.

André, J.C. (2017). Convergence NBIC : risques et conditions de possibilité au regard d'autres risques silencieux. *Environnement risques et santé*, 16, 178–191.

André, J.C. and Masse, R. (2002). L'expertise, la science et l'incertitude : l'expertise scientifique ou la langue d'Ésope. *Environnement, risques et santé*, 1, 299–306.

Andrew, M., Ford, P.R., Miller, M.T., McRobert, A.P., Foster, N.C., Seerden, G., Littlewood, M.A., Hayes, S.J. (2021). Bridging the gap between science and application: The use of co-creation educational workshops in professional youth soccer. *International Sport Coaching Journal*, 9, 1–18.

Appel, K.O. (2001). La relation entre morale, droit et démocratie. La philosophie du droit de Jürgen Habermas jugée du point de vue d'une pragmatique transcendantale. *Les études philosophiques*, 56, 67–80.

Arbib, J., Dorr, A., Seba, T. (2021). RethinkX – Disruption, implications, and choices [Online]. Available at: https://static1.squarespace.com/static/585c3439be65942f022bbf9b/t/6107fd0ed121a02875c1a99f/1627913876225/Rethinking+Implications.pdf.

Arendt, H. (1968). *The Human Condition*. Calmann-Lévy, Paris.

Arendt, H. (1972). *Le système totalitaire*. Le Seuil, Paris.

Argyris, C. (1999). *Savoir pour agir, surmonter les obstacles à l'apprentissage organisationnel*. Interéditions, Paris.

Attali, J. (ed.) (2007). Introduction. In *Un monde en mouvement ; enjeux et défis*, Attali, J. and de Boissieu, C. (eds). ESKA, Paris.

Aubens, F. and Benasayag, M. (1999). *La fabrication de l'information*. La Découverte, Paris.

Audoin, L. (1999). Bienvenue dans l'ère du bienêtre. *L'usine nouvelle*, 2710, 54–60.

Baker, M. (2016). 1,500 scientists lift the lid on reproducibility. *Nature*, 533, 452–454.

Barré, R. (2020). Pour une recherche scientifique responsable. *Natures sciences sociétés*, 28, 45–50.

Baude, O. (ed.) (2006). *Corpus oraux, guide de bonnes pratiques*. CNRS/Presses Universitaires d'Orléans, Orléans.

Bauer, M., Dubois, M., Hervois, P. (2021). Les français et la science 2021 – Représentations sociales de la science 1972–2020. Report [Online]. Available at: http://www.science-and-you.com/sites/science-and-you.com/files/users/documents/les_francais_et_la_science_2021_-_rap-port_de_recherche_web_v29112021_v2.pdf.

Beck, U. (2003). *La société du risque – sur la voie d'une autre modernité*. Flammarion, Paris.

Belasco, W. (2006). *Meals to Come. A History of the Future of Food*. University of California Press, Berkeley.

Bell, A., Chetty, R., Jaravel, X., Petkova, N., Van Reenen, J. (2019). Do tax cuts produce more Einsteins? The impact of financial incentives vs. exposure to innovation on the supply of inventors. CEP Discussion Paper No. 1597 [Online]. Available at: https://cep.lse.ac.uk/pubs/download/dp1597.pdf.

Bensaïd, D. (1999). *Éloge à la résistance à l'air du temps*. Textuel, Paris.

Bensaude-Vincent, B. (2009). *Les vertiges de la technoscience : façonner le monde atome par atome*. La Découverte, Paris.

Bergel, J.L. (1993). Préface. In *Évaluation législative et lois expérimentales*, Morand C.A. (ed.). Presses Universitaires d'Aix-Marseille, Aix en Provence.

Berger, S. (2006). *Made in monde*. Le Seuil, Paris.

Berger-Douce, S. (2015). La performance par l'innovation responsable. *Entreprendre & innover*, 24, 37–44.

Berkun, S. (2007). *The Myths of Innovation*. O'Reilly Media Inc., New York.

Berthelot, J.M., Martin, O., Collinet, C. (2005). *Savoirs et savants ; les études sur la science en France*. PUF, Paris.

Blondiaux, L. (2008). *Le nouvel esprit de la démocratie : actualité de la démocratie participative*. Le Seuil, Paris.

Boissonnat, J. (2007). Dieu et l'argent. *Sociétal*, 57, 105–107.

Boudon, R. (1977). *Effets pervers et ordre social*. PUF, Paris.

Boudon, R. (2009). *La rationalité*. PUF, Paris.

Bouet, B. (2009). Analyse des risques. Master's Thesis, Université de Bordeaux 2, Bordeaux.

Bourdieu, P. (1973). *L'opinion publique n'existe pas !* Les Temps Modernes, Paris.

Bourg, D. and Schlegel, J.L. (2001). *Parer aux risques de demain : le principe de précaution*. Le Seuil, Paris.

Brumefield, J. (2009). Supplanting the old media? *Nature*, 458, 546–558.

Bruner, J. (2002). *Pourquoi nous racontons nous des histoires ?* Retz, Paris.

Bush, V. (1980). *The Endless Frontier: A Report to the President*. Arno Press, New York [1945].

Calame, P. (1999). *Manifeste pour une science citoyenne, responsable et solidaire*. FPH, Paris.

Callon, M. (2003). Science et société : les trois traductions. *Les Cahiers du MURS*, 42, 57–71.

Callon, M., Lascoumes, P., Barthe, Y. (2001). *Agir dans un monde incertain – essai sur la démocratie technique*. Le Seuil, Paris.

Charbonneau, J. and Estèbe, P. (2001). Entre l'engagement et l'obligation : l'appel à la responsabilité à l'ordre du jour. *Lien social et politiques*, 46, 5–15.

Charlot, B. (1995). *Les sciences de l'éducation, un enjeu, un défi*. ESF, Paris.

Chevassus-au-Louis, B. (2007). *L'analyse des risques : l'expert, le décideur et le citoyen*. Quae, Paris.

Chouteau, M., Forest, J., Nguyen, C. (2020). *Innovation for Society*. ISTE Ltd, London, and John Wiley & Sons, New York.

Coller, B.S. and Califf, R.M. (2009). Traversing the valley of death: A guide to assessing prospects for translational success. *Science Translational Medicine*, 1(10), 10cm9.

COMETS (2006). Enjeux éthiques des nanosciences et des nanotechnologies. CNRS, Paris [Online]. Available at: http://www.cnrs.fr/fr/organisme/ethique/publis.htm.

Comte-Sponville, A. (2004). *Le capitalisme est-il moral ?* Albin Michel, Paris.

Constant, B. (1997). *Écrits politiques*. Gallimard, Paris.

Coote, A.M. and Franklin, J. (1999). Negotiating risks to public health-models of participation. In *Risk Communication and Public Health*, Bennett, P. and Calman, K. (eds). Oxford University Press, New York.

Corichi García, A., Pérez García, J.C., Pérez Enzastiga, J.A., Valencia Sandoval, K. (2018). Introduction to disruptive innovation and its effect in some industrial sectors. *Strategy, Technology & Society*, 7, 87–109.

Crawford, M.B. (2016). *The World Beyond Your Head: On Becoming an Individual in an Age of Distraction*. Farrar, Strauss and Giroux, New York.

Cuppen, E., van de Grift, E., Pesch, U. (2019). Reviewing responsible research and innovation: Lessons for a sustainable innovation research agenda? In *Handbook of Sustainable Innovation*, Boons, F. and McMeekin, A. (eds). Edward Elgar Publishing, Cheltenham.

Czajkowski, S.M., Lynch, M.R., Hall, K.L., Stipelman, B.A., Haverkos, L., Perl, H., Scott, M.S., Shirley, M.C. (2016). Transdisciplinary translational behavioral (TDTB) research: Opportunities, barriers, and innovations. *Science Translational Medicine*, 6, 32–43.

Dacheux, E. (2013). Redécouvrir les liens entre science et anarchie pour penser l'indiscipline du chercheur et sa nécessaire responsabilité. In *Le gouvernement des technosciences ; gouverner le progrès et ses dégâts depuis 1945*, Pestre, D. (ed.). La Découverte, Paris.

Daloz, P., Johnson, P., Massart, S., Le Masson, P., Weil, B. (2020). Shaping the unknown with virtual universes – The new fuel for innovation. Global Innovation Index, Cornell University, INSEAD, WIPO [Online]. Available at: https://www.tmci.minesparis.psl.eu/wp-content/uploads/2021/01/4.4.pdf.

Davies, S.R. (2019). An ethics of the system: Talking to scientists about research integrity. *Science & Engineering Ethics*, 25, 1235–1253.

Davis, M. and Laas, K. (2014). "Broader impacts" or "responsible research and innovation"? A comparison of two criteria for funding research in science and engineering. *Science and Engineering Ethics*, 20, 963–983.

Debref, R., Gallaud, D., Temple, L., Temri, L. (2019). L'innovation responsable, dimension stratégique des organisations. *Innovations*, 59, 5–13.

Decarsin, J. (2021). Le "vrai" nom de la post-vérité [Online]. Available at: http://agora.qc.ca/chroniques/le-vrai-nom-de-la-post-verite.

Demortain, D. (1999). Vivons-nous dans une civilisation du risque ? Incertitude, savoir et communication. Report, DS-INRS, Nancy.

Descombes, V. (1989). *Philosophie par gros temps*. Éditions de Minuit, Paris.

Deserti, A., Real, M., Schmittinger, F. (2022). *Co-creation for Responsible Research and Innovation*. Springer, Cham.

Desriaux, F. (1999). Un rapport de l'IGAS sur l'INRS : la leçon de déontologie. *Santé et travail*, 28, 5–6.

DGRI (2013). Options pour renforcer la recherche et l'innovation responsables : rapport du groupe d'experts sur l'état de l'art en Europe en matière de recherche et d'innovation responsables. Report [Online]. Available at: http://ec.europa.eu/research/science-society/document_library/pdf_06/options-for-strengthening_en.pdf.

Diaba-Nuhoho, P. and Amponsah-Offeh, M. (2021). Reproducibility and research integrity: The role of scientists and institutions. *BMC Research Notes*, 14, 451.

Dienel, P.C. and Renn, O. (1995). Planning cells: A gate to "fractal" mediation. In *Fairness and Competence in Citizen Participation*, Renn, O., Webler, T., Wiedemann, P. (eds). Springer, Boston [Online]. Available at: https://fr.scribd.com/document/395313081/Fairness-and-Competence-in-Citizen-Participation.

Dietz, T. (1995). Forewords – Democracy and science. In *Fairness and Competence in Citizen Participation*, Renn, O., Webler, T., Wiedemann, P. (eds). Springer, Boston [Online]. Available at: https://fr.scribd.com/document/395313081/Fairness-and-Competence-in-Citizen-Participation.

Donaldson, T. and Preston, E. (1995). The stakeholder theory of the corporation: Concepts evidence and implication. *The Academy of Management Review*, 20, 65–91.

Dortier, J.F. (2015). Innover… contre les idées reçues. *Les grands dossiers des sciences humaines*, 38, 1.

Dosi, G. (1982). Technological paradigms and technological trajectories: A suggested interpretation of the determinants and directions of technical change. *Research Policy*, 2, 147–162.

Dreyer, M., Chefneux, L., Goldberg, A., Von Heimburg, J., Patrignani, N., Schofield, M., Shilling, C. (2017). Responsible innovation: A complementary view from industry with proposals for bridging different perspectives. *Sustainability*, 9, 1719.

Dufresne, J. (2021). Les trois virus : biologique, informatique et idéologique [Online]. Available at: http://agora.qc.ca/jacques-dufresne/les-trois-virus-biologique-informatique-et-ideo-logique.

Durkheim, E. (1963). *Sociologie et philosophie*. PUF, Paris.

Dutton, W.H. (2004). *Social Transformation in an Information Society: Rethinking Access to You and the World*. UNESCO, Paris.

Edgerton, D. (2010). Innovation, technology, or history: What is the historiography of technology about? *Technology and Culture*, 51, 680–697.

Einstein, A. (1991). *Écrits politiques d'Albert Einstein*. Le Seuil, Paris.

Ellul, J. (1954). *La technique, ou l'enjeu du siècle*. Armand Colin, Paris.

Ellul, J. (1988). *Le bluff technologique*. Hachette, Paris.

Elster, J. (2009). *Le désintéressement : traité critique de l'Homme économique*. Le Seuil, Paris.

Estrades, J. and Rémy, E. (2003). *L'expertise en pratique*. L'Harmattan, Paris.

Etchegoyen, A. (1999). *La vraie morale se moque de la morale – être responsable*. Le Seuil, Paris.

EU (2008). Recommandation de la commission du 7 février 2008 concernant un code de bonne conduite pour une recherche responsable en nanosciences et nanotechnologies. *Journal officiel de l'Union Européenne*, L 116/46-52 2008/345/CE, Brussels.

EU (2019). Pathways declaration. The future of responsible research and innovation (RRI) in "Horizon Europe" [Online]. Available at: http://pathways2019.eu/declaration/.

Everett, H. (2022). Powering the renewable energy transition with 3D printing: Decarbonization [Online]. Available at: https://3dprintingindustry.com/news/powering-the-renewable-energy-transition-with-3d-printing-decarbonization-208029/.

Fanelli, D., Costas, R., Larivière, V. (2015). Misconduct policies, academic culture and career stage, not gender or pressures to publish, affect scientific integrity. *PLoS One*, 10, e0127556.

Feenberg, A. (2004). *(Re)penser la technique : vers une technologie démocratique*. La Découverte/M.A.U.S.S., Paris.

FI (2020). L'investissement ESG devient la nouvelle norme [Online]. Available at: https://www.finance-investissement.com/nouvelles/actualites/linvestissement-esg-devient-la-nouvelle-norme/.

Fischhoff, B. (1998). Risk perception and communication unplugged: Twenty years of process. In *Risk and Modern Society*, Lofstedt, R. and Frewer, L. (eds). Earthscan Pub., London.

Flahaut, F. (2013). Pour une conception renouvelée du bien commun. *Études*, 418, 773–783.

Fløysand, A. and Jakobsen, S.E. (2017). Industrial renewal: Narratives in play in the development of green technologies in the Norwegian salmon farming industry. *The Geographical Journal*, 183, 140–151.

Foray, D. (2001). Science, technologie et marché. In *Les sciences sociales dans le monde*, Bickerstaff, J.C. (ed.). UNESCO/MSH, Paris.

Forman, F. (2007). The primacy of sciences in modernity, of technology in postmodernity, and of ideology in the history of technology. *History and Technology*, 23, 1–152.

Forsyth, T. and Johnson, C. (2014). Elinor Ostrom's legacy: Governing the commons and the rational choice controversy. *Development and Change*, 45, 1093–1110.

Frese, M., Bausch, A., Schmidt, P., Rauch, A., Kabst, R. (2012). Evidence-based entrepreneurship: Cumulative science, action principles, and bridging the gap between science and practice. *Foundations and Trends in Entrepreneurship*, 8, 1–62.

Fukuyama, F. (2002). *La fin de l'Homme : les conséquences de la révolution biotechnique*. Folio-Actuel, Paris.

Gandhi, M.K. (1969). *Tous les Hommes sont frères – vie et pensées du Mahatma Gandhi d'après ses œuvres*. Gallimard-Idées, Paris.

Gauchet, M. (2002). *La démocratie contre elle-même*. Gallimard, Paris.

Gaudin, T. (2005). *La prospective*. PUF, Paris.

Gibbons, M., Limoges, C., Nowotny, H., Schwartmon, S., Scott, P., Trow, M. (1994). *The New Production of Knowledge: The Dynamic of Science and Research in Contemporary Societies*. Sage, London.

Giddens, A. (1994). *Les conséquences de la modernité*. L'Harmattan, Paris.

Girard, R. (1973). *La violence et le sacré*. Pluriel-Hachette, Paris.

Girard, R. (1982). *Le bouc émissaire*. Grasset, Paris.

Girard, R. (2002). *La voie méconnue du réel : une théorie des mythes archaïques et modernes*. Grasset, Paris.

Godard, O. (2007). Est-il légitime de soumettre la connaissance scientifique au débat public. *Réalités industrielles, Annales des Mines*, 12–18.

Godard, O., Henry, C., Lagadec, P., Michel-Kerjan, E. (2002). *Traité des nouveaux risqués*. Gallimard-Folio, Paris.

Gonod, P.F. (1990). Problématique de la maîtrise sociale de la technologie. *Analyse des Systèmes*, XVI [Online]. Available at: https://asrdlf.org/pdf/PDFeve69CN1.pdf.

Grenier, J.Y. and Lepetit, B. (1989). L'expérience historique – à propos de C.E. Labrousse. *Annales ESC*, 6, 1337–1360.

Grundwald, A. (2014). The hermeneutic side of responsible research and innovation. *Journal of Responsible Innovation*, 1, 274–291.

Gurca, A. and Ravishankar, M.N. (2016). A bricolage perspective on technological innovation in emerging markets. *IEEE Transactions on Engineering Management*, 63, 53–66.

Gust-Desprairies, F. (2003). *L'imaginaire collectif*. Erès, Toulouse.

Habermas, J. (2002). *L'avenir de la nature humaine ; vers un eugénisme libéral ?* Gallimard, Paris.

Hacking, J. (2001). *Entre science et réalité ; la construction sociale de quoi ?* La Découverte, Paris.

Halioua, B. (2017). *Le procès des médecins de Nuremberg*. Erès, Toulouse.

d'Haraucourt, P.J. (2021). Nous français, avons la meilleure administration du monde [Online]. Available at: https://www.contrepoints.org/2021/01/18/389074-nous-francais-avons-la-meilleure-administration-du-monde.

Haven, T.L., Bouter, L.M., Smulders, Y.M., Tijdink, J.K. (2019a). Perceived publication pressure in Amsterdam: Survey of all disciplinary fields and academic ranks. *PLoS One*, 14, e0217931.

Haven, T.L., Tijdink, J.K., Martinson, B.C., Bouter, L.M. (2019b). Perceptions of research integrity climate differ between academic ranks and disciplinary fields: Results from a survey among academic researchers in Amsterdam. *PLoS One*, 14, e0210599.

Haven, T.L., Pasman, H.R., Widdershoven, G., Bouter, L., Tijdink, J. (2020). Researchers' perceptions of a responsible research climate: A multi focus group study. *Science and Engineering Ethics*, 26, 3017–3036.

Haven, T.L., Tijdink, J.K., Martinson, B.C., Bouter, L.M., Oort, F.J. (2021). Explaining variance in perceived research misbehavior – Results from a survey among academic researchers in Amsterdam. *Research Integrity and Peer Review*, 6(7).

von Hayek, F. (2022). Dernier entretien de Friedrich Hayek [Online]. Available at: https://www.contrepoints.org/2022/03/13/200605-dernier-entretien-de-friedrich-hayek.

HCSP (2009). La sécurité des travailleurs lors de l'exposition aux nanotubes de carbone [Online]. Available at: https://www.hcsp.fr/explore.cgi/avisrapportsdomaine?clefr=55.

HCSP (2010). Recommandation de vigilance relative à la sécurité des nanoparticules d'argent [Online]. Available at: https://www.hcsp.fr/Explore.cgi/avisrapportsdomaine?clefr=122.

HCSP (2018). Protéger les travailleurs et les personnes au voisinage de sites de production ou de manipulation de nanoparticules de dioxyde de titane [Online]. Available at: https://www.hcsp.fr/Explore.cgi/avisrapportsdomaine?clefr=663.

Heidegger, M. (1927). *Sein und Zeit*. Niemeyer, Tübingen.

Heidegger, M. (1958). Que veut dire penser ? In *Essais et conférences*. Gallimard, Paris.

Heidegger, M. (2003). *La question de la technique*. Gallimard, Paris.

Heiderich, D. (2007). Éditorial. *Magazine de la communication de crise sensible*, 15, 2–3.

Hirsch, E. (2021). L'approche éthique, une exigence politique. In *De nouveaux territoires pour l'éthique de la recherche – repères, responsabilités et enjeux*, Coutellec, L., Coulée, F., Hirsch, E. (eds). Erès, Toulouse.

Horbach, S.P.J.M. and Halffman, W. (2019). The extent and causes of academic text recycling or "self-plagiarism". *Research Policy*, 48, 492–502.

Horizon 2020 (2020). Charte européenne du chercheur. Code de conduite pour le recrutement des chercheurs [Online]. Available at: https://www.horizon2020.gouv.fr/cid96323/charte-europeenne-chercheur-code-conduite-pour-recrutement-des-chercheurs.html.

Horizon Europe (2019). The next EU Research & Innovation Programme (2021–2027) [Online]. Available at: https://prod5.assets-cdn.io/event/4336/assets/8428823124-12a7724d30.pdf.

House of Lords (2000). Science and society, report of the select committee on science and technology [Online]. Available at: http://www.parliament.the-stationery-office.co.uk/pa/ld199900/ldselect/ldsctech/38/3802.htm.

Hume, D. (1975). *Œuvres complètes*. Gallimard, Paris.

Ingham, M. (2011). *Vers l'innovation responsable : pour une vraie responsabilité sociétale*. De Boeck, Brussels.

Irwin, A. (1995). *Citizen Science – A Study of People, Expertise and Sustainable Development*. Routeledge, London.

Jacob, F. (1997). *La souris, la mouche et l'Homme*. Odile Jacob, Paris.

Jaeger, M. and Mispelblom-Beyer, F. (2014). Introduction. In *Le travail social et la recherche – conférence de consensus*, Jaeger, M. (ed.). Dunod, Paris.

Jakobsen, S.E., Fløysand, A., Overton, J. (2019). Expanding the field of responsible research and innovation (RRI) – From responsible research to responsible innovation. *European Planning Studies*, 27, 2329–2343.

Janicaud, D. (2002). *L'Homme va-t-il dépasser l'humain ?* Bayard, Paris.

Jaulin, R. (1970). *La paix blanche : introduction à l'ethnocide*. Collection 10/18, Paris.

Jaulin, R. (1977). *Les chemins du vide*. Collection 10/18, Paris.

John, L.K., Loewenstein, G., Prelec, D. (2012). Measuring the prevalence of questionable research practices with incentives for truth telling. *Psychological Science*, 23, 524–532.

Jonas, H. (1990). *Le principe responsabilité : une éthique pour la civilisation technologique*. Le Cerf, Paris.

Julliard, J. (2008). *La reine du monde*. Flammarion, Paris.

Jurdant, B. (1994). Le contrôle social de la science. In *Science ou justice*, Heilman, E. (ed.). Autrement, Paris.

Kammoun, S. and Le Bas, C. (2020). L'innovation frugale peut-elle être disruptive. In *La revanche des communs – science et créativité*. De Boeck Supérieur, Paris.

Kant, E. and Mendelsohn, M. (2007). *Qu'est-ce que les Lumières*. Mille et une nuits, Paris.

Kaufmann, J.C. (2001). *Ego : pour une sociologie de l'individu*. Nathan, Paris.

Kaufmann, J.C. (2008). *Quand Je est un autre – pourquoi et comment ça change en nous ?* Armand Colin, Paris.

Kerninon, Y. (2009). *Tentative d'assassinat du bourgeois qui est en moi*. Méta-Éditions, Paris.

Keynes, J.M. (1953). *The General Theory of Employment, Interest, and Money*. Harcourt Jovanovich, New York.

Kijkuit, B. and Van Den Ende, J. (2007). The organizational life of an idea: Integrating social network, creativity and decision-making perspectives. *Journal of Management Studies*, 44, 863–882.

Klein, E. (2008). *Galilée et les Indiens*. Flammarion, Paris.

Knorr-Cetina, K. (1999). *Epistemic Cultures: How the Sciences Make Knowledge*. Harvard University Press, Boston.

Koslowki, P. (1998). *Principe d'économie éthique*. Le Cerf, Paris.

Koyré, A. (1968). *From the Closed World to the Infinite Universe*. Gallimard, Paris.

Kreps, D.M., Milgrom, P., Roberts, J., Wilson, R. (1982). Rational cooperation in the finitely repeated prisoner's dilemma. *Journal of Economic Theory*, 27, 245–252.

Kuhn, T. (1983). *La structure des révolutions scientifiques*. Flammarion, Paris.

Laffont, J.J. (1975). Macroeconomic constraints, economic efficiency and ethics: An introduction to Kantian economics. *Economica*, 42, 430–437.

Latour, B. (2018). The post-truth philosopher, mounts a defense of Science. *New York Times* [Online]. Available at: https://www.nytimes.com/2018/10/25/magazine/bruno-latour-post-truth-philosopher-science.html.

Lavelle, S. (2007). Quand connaître, c'est agir. In *Éthique, technique et démocratie*, Fetz, B., Goujon, P., Hériard-Dubreuil, B., Lesch, W. (eds). Bruylant-Academia, Louvain-la-Neuve.

Layet, M. (2008). Wi-Fi : bluesafe anticipe l'impact sanitaire [Online]. Available at: http://www.novethic.fr:80/novethic/v3/article.jsp?id=117827.

Leblanc, R. (2020). L'asphyxie de l'économie n'est pas un sauvetage [Online]. Available at: https://www.institut-entreprise.fr/robert-leblanc-pdg-daon-france-et-vice-president-de-la-region-emea-daon-risk-solutions.

Lechevalier, S. (ed.) (2019). *Innovation Beyond Technology: Science for Society and Interdisciplinary Approaches*. Springer, Singapore.

Lehoux, P., Daudelin, G., Denis, J.L., Gauthier, P., Hagemeister, N. (2019). Pourquoi et comment sont conçues les innovations responsables ? Résultats d'une méta-ethnographie. *Innovations*, 59, 15–42.

Lénel, P. (2014). Des dissensus au consensus : point de fuite et lignes de convergence. In *Le travail social et la recherche – conférence de consensus*, Jaeger, M. (ed.). Dunod, Paris.

Leprince-Ringuet, L. (1980). *La potion magique*. Flammarion, Paris.

Leroi-Gourhan, A. (1965). *Le geste et la parole II. La mémoire et les rythmes*. Albin Michel, Paris.

Leydesdorff, L. and Etzowitz, H. (2000). Le mode 2 et la globalisation des systèmes d'innovation natio-naux : le modèle en triple hélice des relations entre universités, industrie et gouvernement. *Sociologie et sociétés*, 32, 135–156.

Li Vigni, F. (2021). Cinq types de travail scientifique "interdisciplinaire". *Natures sciences sociétés*, 29, 130–140.

Libaert, T. (2003). *La transparence en trompe l'œil*. Descart & Cie, Paris.

Lipovetsky, G. (1992). *Le crépuscule du devoir : l'éthique indolore des nouveaux temps démocratiques*. NRF Essais, Paris.

Lipovetsky, G. and Serroy, J. (2008). *La culture-monde : réponse à une société désorientée*. Odile Jacob, Paris.

Maalouf, A. (1998). *Les identités meurtrières*. Livre de poche, Paris.

Malka, S. and Gregori, M. (2005). *Infiltration : une taupe à la solde de Philip Morris*. Georg, Lausanne.

Marchard, P. (2022). Covid : 157 articles scientifiques ont été rétractés depuis le début de la pandémie [Online]. Available at: https://www.egora.fr/actus-pro/recherche/71333-covid-157-articles-scientifiques-ont-ete-retractes-depuis-le-debut-de-la.

Marinese, V. (2008). Comprendre la relation entre la loi, les sciences et le citoyen [Online]. Available at: http://www.lemensuel.net/Comprendre-la-relation-entre-la.html.

Mariotti, H. (2008). Réductionnisme, "holisme", approche systémique et pensée complexe [Online]. Available at: http://www.geocities.com/plurivesu/redufr.html.

Mercier, A. (2008). Repérage de la communication scientifique. In *La communication politique*, Mercier, A. (ed.). CNRS, Paris.

Métais, A., André, J.C., Redlingshoefer, B. (2014). La démocratie à l'épreuve du débat public CIGEO ; mise en scène d'un débat et d'un choix politique. *Droit de l'environnement*, 224, 1–11.

Michéa, J.C. (2006). *L'enseignement de l'ignorance et ses conditions modernes*. Climats, Paris.

Minguet, G. and Osty, F. (2008). *En quête d'innovation : du projet au produit de haute technologie*. Hermes-Lavoisier, Paris.

de la Miranda, L. (2010). *L'art d'être libres au temps des automates*. Max Milo, Paris.

Morel, C. (2002). *Les décisions absurdes*. Gallimard, Paris.

Morin, E. (2000). *Les sept savoirs nécessaires à l'éducation du futur*. Le Seuil, Paris.

Morin, E. (2004). *La Méthode – éthique*. Le Seuil, Paris.

Nazarko, L. (2020). Responsible research and innovation in enterprises: Benefits, barriers and the problem of assessment. *Journal of Open Innovation*, 6(12).

Neubauer, C. and Joly, P.B. (2004). De la culture scientifique à la citoyenneté scientifique. Rapport pour la mission de la "culture et information scientifiques et techniques et des musées" du Ministère de la Recherche. Report, Direction de la Recherche, INRA, Paris.

Nisbet, M.C. (2009a). Framing science: A new paradigm in public engagement. In *Understanding Science: New Agenda in Science communication*, Kahlor, K. and Stout, P. (eds). Taylor & Francis, New York.

Nisbet, M.C. (2009b). The ethics of framing science. In *Communicating Biological Sciences: Ethical and Metaphorical Dimensions*, Nerlich, B., Larson, B., Elliott, R. (eds). Ashgate, London.

Nisbet, M.C. and Scheufele, M. (2009). What's next for science communication? Promising directions and lingering distractions. *American Journal of Botany*, 96, 1767–1778.

Nordon, D. (2006). Méfions-nous du savoir ! [Online]. Available at: http://www.lemensuel.net/Mefions-nous-du-savoir.html.

Nosek B.A., Alter G., Banks G.C., Borsboom D., Bowman S.D., Breckler S.J., Buck S., Chambers C.D., Chin G., Christensen G. (2015). Promoting an open research culture. *Science*, 348, 1422–1425.

Nowotny, H., Scott, P., Gibbons, M. (2003). *Repenser la Science*. Belin, Paris.

Nowotny, H., Pestre, D., Schmidt-Assmann, H., Schultze-Fielitz, H., Trute, H.H. (2005). *The Public Nature of Science under Assault: Politics, Markets, Science and the Law*. Springer, Heidelberg.

Ogien, R. (2007). *L'éthique aujourd'hui ; maximalistes et minimalistes*. Folio essais, Paris.

Omnès, R. (1994). *Philosophie de la science contemporaine*. Folio Essais, Paris.

Owen, R. and Pansera, M. (2019). Responsible Innovation and Responsible Research and Innovation. In *Handbook of Research on Public Policy*, Simon, D., Kuhlmann, S., Stamm, J., Canzler, W. (eds). Edward Elgar Publishing, Cheltenham.

Owen, R., Bessant, J., Heintz, M. (eds) (2013). *Responsible Innovation: Managing the Responsible Emergence of Science and Innovation in Society*. John Wiley & Sons, London.

Pavie, X. (2012). *Innovation responsable – stratégie et levier de croissance des organisations*. Eyrolles, Paris.

Peyrat, D. (2003). *Éloge de la sécurité*. Gallimard, Paris.

Pfeffer, J. and Sutton, R.I. (2000). *The Knowing-Doing Gap*. Harvard Business School Press, Boston.

Philip, P. (2007). Intelligence économique et gestion des risques environnementaux : l'exemple de la protection des travailleurs du nucléaire. In *Intelligence économique et gestion des risques*, Bouchet, H.H. and Guilhon-Le Fraper du Hellen, A. (eds). Pearson Education France, Paris.

Pineo, H., Turnbull, E.R., Davies, M., Rowson, M., Hayward, A.C., Hart, G., Johnson, A.M., Aldridge, R.W. (2021). A new transdisciplinary research model to investigate and improve the health of the public. *Health Promotion International*, 36, 481–492.

Pinsart, M.G. (2009). *La bioéthique*. Le Cavalier Bleu, Paris.

Potter, V.R. (1971). *Bioethics – A Bridge to the Future*. Prentice Hall, Upper Saddle River.

de Pracontal, M. (2005). *L'imposture scientifique en dix leçons*. Le Seuil, Paris.

Prieur, C., Daran, E., Peaucelle, D. (2004). États généraux de la recherche au LAAS-CNRS – Thème Recherche et Société [Online]. Available at: http://www.Articles.php3?id_article=252&var_recherche=pdf.

Puech, M. (2008). *Homosapiens technologicus : philosophie de la technologie contemporaine*. Le Pommier, Paris.

Quéré, L. (2001). La structure cognitive et normative de la confiance. In *La confiance*, Quéré, L. (ed.). Hermès/Réseaux, Paris.

Ramonet, I. (1999). *L'agonie de la culture*. Folio-Actuel, Paris.

Rask, M., Mačiukaitė-Žvinienė, M., Tauginienė, L., Dikčius, V., Matschoss, K., Aarrevaara, T., d'Andrea, L. (2018). *Public Participation, Science and Society – Tools for Dynamic and Responsible Governance of Research and Innovation*. Taylor & Francis Group/Routledge, Abingdon.

Redlingshoefer, B., Métais, A., André, J.C. (2014). Complexité, risques incertains, précaution et débat public. *Environnement, risques et santé*, 13, 222–231.

Reschler, N. (1998). *Complexity: A Philosophical Overview*. Transactions Publishers, New Brunswick.

Resweber, J.P. (2000). *Le Pari de la transdisciplinarité, vers l'intégration des savoirs*. L'Harmattan, Montreal.

Revel, J.F. (1988). *La connaissance inutile*. Grasset, Paris.

Reversat, N. and André, J.C. (2022). Covid19 and process engineering (eBook Edition). Entropie [Online]. Available at: https://www.openscience.fr/Covid19-et-Genie-des-Procedes.

Reynaud, J.D. (1989). *Les règles du jeu*. Armand Colin, Paris.

RF (2008). Décret no. 2008-632 portant sur la "création d'un traitement automatisé de données à caractère personnel dénommé Edvige". Paris.

Richards, R. (1993). Birth, death, and resurrection of evolutionary ethics. In *Evolutionary Ethics*, Nitecki, H. (ed.). Suny Press, Albany.

Richmond, J. (2000). The 3Rs – Past, present and future. *Scandinavian Journal of Laboratory Animal Science*, 27, 84–92.

Ricœur, P. (1990). *Soimême comme un autre*. Le Seuil, Paris.

Roman, J. (1998). *La démocratie des individus*. Calmann-Lévy, Paris.

Ronsavallon, P. (2006). *La contre-démocratie*. Le Seuil, Paris.

Roqueplo, P. (1983). *Penser la technique. Pour une démocratie concrète*. Le Seuil, Paris.

Roqueplo, P. (1996). *Entre savoir et décision, l'expertise scientifique*. INRA, Paris.

Rosa, H. (2020). Le miracle et le monstre – un regard sociologique sur le Coronavirus [Online]. Available at: https://aoc.media/analyse/2020/04/07/le-miracle-et-le-monstre-un-regard-sociologique-sur-le-coronavirus/.

Rousseau, J.J. (2009). *Émile ou de l'éducation*. Garnier-Flammarion, Paris.

Rousseau, D.M. and McCarthy, S. (2007). Educating managers from an evidence-based perspective. *Academy of Management Learning and Education*, 6, 84–101.

Runu, R. (2021). Responsible research. *Journal of Orthopedic Diseases and Traumatology*, 4(59).

Rynes, S.L., Giluk, T.L., Brown, K.G. (2007). The very separate worlds of academic and practitioner periodicals in human resource management: Implications for evidence/based management. *Academy of Management Journal*, 50, 987–1008.

de Saint-Exupéry, A. (1956). Lettre au Général X. In *Un sens à la vie*. Gallimard, Paris.

de Saint-Simon, H. (1923). *Œuvres*. Les grands écrivains français, Paris.

Sauvé, M.R. (2008). Trois produits chimiques seraient inoffensifs chez l'humain… mais cancérigènes chez les rats [Online]. Available at: http://nouvelles.umontreal.ca:80/content/view/883/1/.

Schaer, E., Frochot, C., André, J.C. (2020). Il ne faut pas avoir peur des "nano", c'est juste une nanocrise mal anticipée. *Environnement risques et santé*, 19, 446–466.

Scheufele, D., Corley, E., Shih, T., Dalrymple, K., Ho, S. (2009). Religious belief and public attitudes to nanotechnology in Europe and the United States. *Nature Nanotechnology*, 4, 91–94.

von Schomberg, L. and Blok, V. (2021). The turbulent age of innovation. *Synthese*, 198, 4667–4683.

Schumacher, E.F. (1973a). *Small is Beautiful: A Study of Economics as if People Mattered*. Le Seuil, Paris.

Schumacher, E.F. (1973b). *Small is beautiful : une société à la mesure de l'Homme*. Le Seuil Paris.

Schwarz, M. and Thompson, M. (1990). *Divided We Stand: Redefining Politics, Technology and Social Choice*. University of Pennsylvania Press, Philadelphia.

Serres, M. (2007). Discours de réception à l'Académie Française (1991). In *Le coût de l'excellence*, Aubert, N. and De Gaulejac, V. (eds). Le Seuil, Paris.

Serres, M. (2008). *Le mal propre : polluer pour s'approprier*. Le Pommier, Paris.

Serres, M. (2015). Fenêtres sur cours. In *Dossier Disciplines et interdisciplinarité – les disciplines au cœur de la polyvalence à l'école primaire*, Becker, A. and Garnier, P. (eds). Institut de recherches de la FSU, Paris.

Sfez, L. (1992). *Critique de la communication*. Le Seuil, Paris.

Shapin, S. (1998). *La révolution scientifique*. Flammarion, Paris.

Shin, C. and Park, J. (2019). How social entrepreneurs' value orientation affects the performance of social enterprises in Korea: The mediating effect of social entrepreneurship. *Sustainability*, 15(5341).

Simon, H.A. (1947). *Administrative Behavior*. Macmillan, New York.

Sjøtun, S. (2019). A ferry making waves: A demonstration project "doing" institutional work in a green maritime industry. *Norwegian Journal of Geography*, 73, 16–28.

Slama, A.G. (1993). *L'angélisme exterminateur*. Grasset, Paris.

Sloterdijk, P. (2000). *Critique de la raison cynique*. C. Bourgeois, Paris.

Stengers, I. and Bensaude-Vincent, B. (2003). *100 mots pour commencer à penser les sciences*. Les empêcheurs de penser en rond, Paris.

Stilgoe, J., Owen, R., Macnaghten, P. (2013). Developing a framework for responsible innovation. *Research Policy*, 42, 1568–1580.

Taalbi, J. (2017). What drives innovation? Evidence from economic history. *Research Policy*, 46, 1437–1453.

Tait, J. (2017). From responsible research to responsible innovation: Challenges in implementation. *Engineering Biology*, 1, 7–11.

Taylor, M.C. (2001). *The Moment of Complexity: Emerging Network Culture*. University of Chicago Press, Chicago.

Temple, L., Gaunand, A., Trouche, G., Vall, E. (2018). Évaluer les impacts des recherches en agriculture sur la société et les écosystèmes : outils, méthodes, études de cas. *Cahiers de l'agriculture*, 27(34002).

Ten Have, H. (ed.). (2007). *Nanotechnologies, éthique et politique*. UNESCO, Paris.

Thapa, R.K., Iakovleva, T., Foss, L. (2019). Responsible research and innovation: A systematic review of the literature and its applications to regional studies. *European Planning Studies*, 27, 2470–2490.

Theys, J. and Cornu, P. (2019). Temps de la nature, temps de la société, temps scientifique à l'heure du changement global. Pour une approche interdisciplinaire de la crise de la temporalité. *Natures sciences sociétés*, 27, 381–389.

Thuillier, P. (1972). *Jeux et enjeux de la science – essai d'épistémologie critique*. R. Laffont, Paris.

Tijdink, J.K., Horbach, S.P.J.M., Nuijten, M.B., O'Neill, G. (2021). Towards a research agenda for promoting responsible research practices. *Journal of Empirical Research on Human Research Ethics*, 16, 450–460.

Timmermans, J. (2017). Mapping the RRI landscape: An overview of organizations, projects, persons, areas and topics. In *Responsible Innovation 3. A European Agenda?*, Asveld, L., van Dam-Mieras, R., Swierstra, T., Lavrijssen, S., Linse, K., van den Hoven, J. (eds). Springer, Cham.

de Tocqueville, A. (1981). *Democracy in America*. Flammarion, Paris.

Touraine, A. (1988). *La parole et le sang*. Odile Jacob, Paris.

Tsui, A.S. (2021). Responsible research and responsible leadership studies. *Academy of Management Discoveries*, 7, 166–170.

Tubiana, M. (2007). *N'oublions pas demain ; mémoires*. Éditions de Fallois, Paris.

UNESCO (2000). *Science for the Twenty-first Century: A New Commitment*. UNESCO, Paris.

UNESCO (2005). Déclaration universelle sur la bioéthique et les droits de l'Homme [Online]. Available at: https://www.unesco.org/fr/ethics-science-technology/bioethics-and-human-rights.

UNESCO (2021). Understanding the impact of artificial intelligence on skills development [Online]. Available at: https://unesdoc.unesco.org/ark:/48223/pf0000376162.

Vallayes, F. (2020). Misère du gouvernement par la peur et la disproportion [Online]. Available at: https://up-magazine.info/decryptages/analyses/48526-misere-du-gouvernement-par-la-peur-et-la-disproportion/.

Van de Vliert, E. and Murray, D.R. (2018). Climate and creativity: Cold and heat trigger invention and innovation in richer populations. *Creativity Research Journal*, 30, 17–28.

Van Stigt, R., Driessen, P.P.J., Spit, T.J.M. (2015). A user perspective on the gap between science and decision-making. Local administrators' views on expert knowledge in urban planning. *Environmental Science & Policy*, 47, 167–176.

Véron, J.P. (1979). *Dictionnaire des idées non reçues*. Erès, Toulouse.

Veyne, P. (1983). *Les Grecs ont-ils cru à leurs mythes ?* Le Seuil, Paris.

Vinck, D. (1995). *Sociologie des sciences*. Armand Colin, Paris.

Virilio, P. (1996). *Un paysage d'événements*. Galilée, Paris.

Virilio, P. (2001). *Cybermonde, la politique du pire*. Textuel, Paris.

Voltaire, F.M. (2009). *Le philosophe ignorant*. Le livre de Poche, Paris.

Walecka-Jankowska, K. and Zimmer, J. (2019). Open innovation in the context of organizational strategy. *Engineering Management in Production and Services*, 11, 86–95.

Weber, M. (1985). *L'éthique protestante et l'esprit du capitalisme*. Plon, Paris.

Weil, S. (1955). *Oppression and Liberty*. Gallimard, Paris.

Whyte, W.F., Greenwood, D.J., Lazes, P. (1989). Participatory action research: Through practice to science in social research. *American Behavioral Scientist*, 32, 513–551.

Wikipedia (2022). Responsabilité sociale des entreprises [Online]. Available at: http://fr.wikipedia.org/wiki/responsabilite_sociale_des_entreprises.

Wittrock, C., Forsberg, E.M., Pols, A., Macnaghten, P., Ludwig, D. (2021). *Implementing Responsible Research and Innovation: Organizational and National Conditions*. Springer International Publishing, Cham.

Yun, J., Egbetoku, A., Zhao, X. (2019). How does a social open innovation succeed? Learning from burro battery and grassroots innovation festival of India. *Science, Technology and Society*, 24, 122–143.

Zweiacker, P. (2007). *Morts pour la science*. Presses Polytechniques et Universitaires Romandes, Lausanne.

2

3D, 4D and Bio-printing Innovations and Additive Manufacturing

Innovation can be made up of two partially interdependent parts, as we have shown:

– the generation of an idea transformed into an invention which is more technological, translated into a proof of concept (POC);

– the conversion of this invention into a profitable application via an industrial demonstrator (ID); this includes social characteristics (desirability, economics, transfer and training, dissemination, risks, liability, etc.).

Based on commercial success, technological innovation is proactively directed toward organization and the orientation of human and financial resources toward forms of efficiency. It then leads to:

– the need to create new knowledge to support incremental innovations; generation of technical ideas that complement or build on the founding concept created, to find or improve products, manufacturing and services (idea of incremental technological advances);

– the development of these increments (some of which are disruptive) in POC and then DI in the form of a functional prototype; their transfer to manufacturing, distribution and use;

– the emergence of a new disruptive concept.

For a color version of all the figures in this chapter, see www.iste.co.uk/andre/knowledge2.zip.

In light of additive manufacturing processes (with which the author is quite familiar, at least in an academic context), this chapter looks at how this knowledge and the associated innovations relate to these considerations, in particular by taking into account the effects of the 12 "death valleys" defined in Volume 1 and Chapter 1 of Volume 2.

> Contrary to what I might have thought, the scientific approach was not simply a matter of observing, accumulating experimental data and deriving a theory. It began with the invention of a possible world, or fragment of a possible world, to confront it, through experimentation, with the outside world. And it was this endless dialogue between imagination and experience that enabled us to form an ever-finer representation of what we call 'reality' (Jacob 1987).

> In fact, the researcher adopts both positions: the conceptualist often has to correct his/her hypotheses following his empirical (mis)adventures, and, conversely, the empiricist is obliged to start from a hypothesis, however sketchy, in order to undertake his experiments (Chapouthier 2008).

> Scientific discovery is impossible without faith in ideas which are of a purely speculative kind, and sometimes even quite hazy; a faith which is completely unwarranted from the point of view of science, and which, to that extent, is 'metaphysical' (Popper 1989).

2.1. Introduction

> As long as his problem is concerned, the engineer does not participate in a social process in which others can make independent decisions, but lives in a separate world of his own. […] He doesn't have to find out what resources are available, or the relative importance of different needs. [...] He has been given knowledge about the ownership of things which changes nowhere and at no time, and which is independent of any particular human situation (von Hayek 1953).

There are implicit biases against the development of creativity for fear of its disruptive and challenging nature. Creativity means change without the certainty of desired and desirable results. We therefore accept the possibility of failure as a partner in creativity. Indeed, we need to go further than the simple idea, if only to achieve a POC, or even a demonstrator. If Dortier (2015) is right when he writes:

> Are innovations revolutionary? [Probably less so than invention], are they based on originality, the surprising, the unexpected? On the contrary, most are very ordinary. They are based on simple, predictable laws of evolution that nature itself knows how to put to good use. To innovate is [just] to think outside the box. Then it's just a matter of arranging known knowledge.

Then, very quickly, the transcription of the idea, which at the outset could have many freedoms and possible detours, will become a less flexible, adaptable, modifiable construct, until it is set in stone. And financially, it is at the industrialization stage that the risk is greatest, since inflexible production methods have to be envisaged, which means a financial risk that needs to be kept under control:

> Encouraging the emergence of innovative ecosystems means more than just influencing specialization, increasing strategic autonomy. It's another way of looking at security from a dynamic perspective. In research, more than anywhere else, creative destruction is at work. An effective strategy for industrial recovery will have to address the general and specific factors behind France's loss of competitiveness (Aghion et al. 2021).

3D printing is an emerging technology. Despite its many advantages, it has taken almost 20 years for it to become a credible technology. Today, science, technology and the economic market are developing rapidly (around 20%/year) for a variety of reasons which will be analyzed below; but, in reality, this technology has found its place in the economic system. In the early 2010s, 3D printing was followed by other forms of manufacturing, such as 4D printing and bio-printing. The latter can simply be described as the 4D printing of the living.

Unlike "conventional" 3D additive manufacturing, the idea is to play with (at least) one adaptive material which can see the shape or functionality of the 3D object evolve under the influence of a localized or non-localized energy stimulus. We therefore have a large number of degrees of freedom at our disposal, allowing, in principle, less targeted applications than "simple" 3D! Figure 2.1 gives an indication of the temporal evolution of scientific work on 4D printing and bio-printing. These two fields open up extraordinary levels of possibility, such as the possibility of creating objects that can attempt to replace current mechanical systems, material transformations of matter and energy, prostheses, etc., for the former, and biological tissues or organs, and organoids (tissue mimetics for drug testing and precision medicine (André 2021), etc.) for the other. As you may expect, these two fields offer applications with a much wider spectrum of uses than 3D printing.

Therefore, based on the author's experience in the field, and in the light of what has been written about the creativity-application transition, we will examine the elements that facilitate these science-societies (or not).

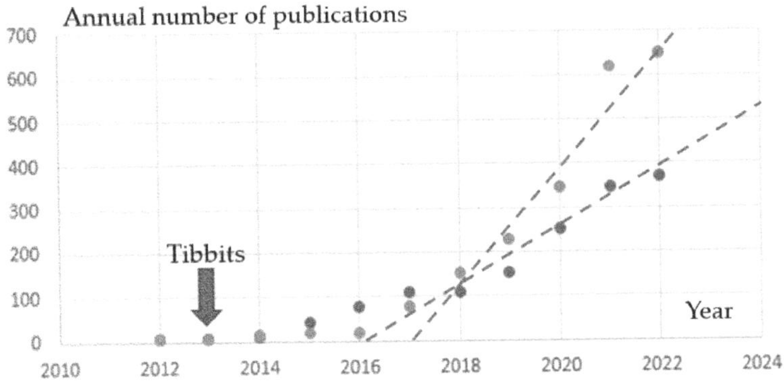

Figure 2.1. *Increase in the number of annual publications: orange: 4D printing; blue: bio-printing (source: CNRS website)*

2.2. Additive manufacturing or 3D printing

> The innovator is in a state of mind of constructive dissatisfaction: he's/She's never satisfied with the best he's got, he's constantly on the lookout for what will enable him to go even further in satisfying market expectations (Libmann et al. 2011).

> Multicultural, globalization must be polytechnic. But this techno-scientific diversification is by no means assured: it depends on the economic and political powers likely to encourage or limit it, on both the supply and demand sides (Hottois 2009).

> Why is philosophy important to engineering? Ultimately and most deeply it is because engineering is philosophy – and through philosophy engineering will become more itself (Mitcham 1998).

Rodrik (2009) has shown that, since the 1960s, economic growth has been strongly associated with the development of modern industrial sectors such as manufacturing. However, after losing 41% of its manufacturing jobs in 30 years, the United States is among the most active players in developing a new, manufacturing-oriented vision. They consider that manufacturing remains a crucial source of high-quality jobs; the production of tradable manufactured goods helps to maintain the balance of trade, and this industry remains an engine of economic growth, because of its higher productivity and opportunities for innovation (Andreoni and Gregory 2013). In this context, radical technologies can be included in technological revolutions (see industry 4.0 and André (2019)); that is, a "process of creative destruction" which creates new sectors and profoundly impacts existing industries, destroying some and rejuvenating others.

Imagination is a way of transforming the forms of desires, daydreams and even intimate fantasies into images; according to Alapetite (2014), it produces the "fluid" basic material for creation, which corresponds to the ability to transform this "material" into signs that can be communicated to others. Sometimes, it does not take much to stimulate it, as Alain Le Méhauté did in 1984 to give rise to the concept of additive manufacturing (and the first patent on the subject; see André et al. (1984)). When these still vague, unstructured elements are linked together in rational sequences, repetition creates content and becomes an organizing element, linking separate, apparently unrelated elements (Einstein 1991). Rouquette (2007) defines creativity as "the ability of an individual, in problem-solving processes, to find novel relationships, to produce new ideas and original solutions in a relatively fluent and flexible way". He adds: "It's not intrinsic value that underpins legitimacy, it's the other way around. Consequently, the question of creativity is not the recognition of value, but the value of recognition".

As already discussed, a common thread running through creativity lies in the classic descriptions of the theme by Guilford (1956), with the aforementioned aspects of divergence and convergence. The designer identifies options in what is readily available to them, exploring a range of solutions and information drawn from their knowledge. Then, as far as possible, they will distinguish between these solutions, without delving too deeply into them, but with common sense. It is on this basis of evaluation (as a rule of thumb) that they will be able to select a single result appropriate to the question posed in this chapter. As can be seen, at this stage, it is just a rough draft that will enable you to tackle the next steps, which are aimed at invention, and then innovation. And while this "rough draft" may lead to the satisfaction of societal needs, self-evaluation and the associated self-satisfaction are no guarantee of an optimal choice! Imagination concerning the same subject, but handled by another person, has no reason to be an identical copy of the initial creativity. On the contrary, paradoxically, it should be a source of variety, because its authors have different cultures and desires. It is the diversity of contexts and environments and the heterogeneity of designers that would prevent the uniform convergence of invention processes. Between small and large inventions, Wilson et al. (1953) consider that an idea is all the more original the rarer it is in the population considered, the more ingenious it is, or the more it combines elements of knowledge that are more remote.

This introductory section poses two questions: the first is a solution, at the origin of additive manufacturing, and the second is linked to other ways of producing objects in 3D.

2.2.1. *Invention of the additive manufacturing concept*

> For one definition of engineering is: the use of heuristics to bring about the best change in a poorly understood situation, within the limits of available resources. For engineering, however advanced it may be, like technology, has the disadvantage of its advantage: it is originally about action, not knowledge. It will always be under the pressure of time, circumstances and limited resources, it is in a way a digest of science in a world of scarce resources, in a limited time and under the pressure of strong demands (Dubois and Brault 2021).

> 'Commitment to science' is therefore a twofold act: i) ontological commitment: the choice of a conceptual framework for the exploration of reality; ii) axiological commitment: the reasons for choosing science as the exclusive means of attaining knowledge of the real – by virtue of the decree of reason enjoining us to consider the operation and methods of science as the most adequate instruments for the project of giving an account of the world – and, moreover to account for the world – and, consequently, as a major goal of culture (Silberstein 2013).

> Our civilization struggles to extricate itself from the fascination of the past. It only dreams about the future, and when it develops projects that are no longer mere dreams, it draws them on a canvas where it is still the past that is projected. It is stubbornly retrospective. It needs to become 'prospective' (Goux-Baudiment 2002).

Gregor and Hevner (2013) consider the existence of four typical invention configurations:

– design which applies known solutions to known problems;

– extension of known solutions to a new problem;

– improvement based on new solutions for known problems;

– invention which consists of creating new solutions to new problems.

We are thus moving from incremental research to something more disruptive. When Alain Le Méhauté asks his question, the fourth configuration involves creating an object using two lasers whose positioning is computer-controlled. You have to invent almost everything! But a technology can be built around the reliable exploitation of any effect, as envisaged by a principle of use. Additive manufacturing, as we shall see, has its own concept, purpose, combination of known components, an architecture, and embodies a generic principle that exploits a basic phenomenon. "We could therefore define a new technology as corresponding to a significant change in one of these areas" (Arthur 2005):

> Excited by his/her project, his/her first practical step is nonetheless retrospective: he/her has to turn to an already existing set of tools and materials; make or re-make an inventory of them; and last but not least, engage in a kind of dialogue with them, to list, before choosing between them, the possible answers that the set can offer to the problem he poses. All these heterogeneous objects constitute his/her treasure trove, and he/her questions them to understand what each of them might 'signify', thus helping to define a whole to be realized, but which will ultimately differ from the instrumental whole only in the internal arrangement of its/her parts (Lévi-Strauss 1990).

Specializing in the study of "simpler" interactions between light and matter, the author, engaged in research into photochemical engineering (or industrial photochemistry), naturally thinks in terms of one-photon light absorption (unlike Battelle's patents, which served as a starting point for Alain Le Méhauté's question); this leads them to think in terms of optical thickness μ (the thickness corresponding to the light crossing with an output transmission of 1/e). Figure 2.2 shows the meaning of μ:

$$\mu = 1/(\varepsilon.c)$$

where ε is a coefficient dependent on the absorbing agent, its concentration and wavelength (called the molecular extinction coefficient), and c is the concentration of the absorbing agent.

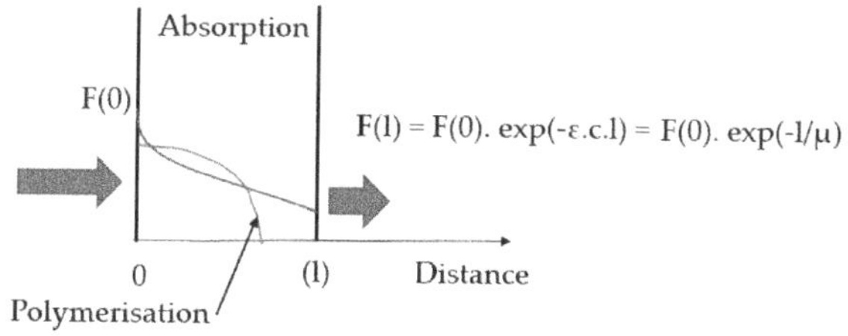

Figure 2.2. *Optical thickness μ*

In transparent media, monophotonic absorption follows Beer–Lambert's law. Locally, for a given concentration c, the absorbed light intensity is proportional to the product c. exp(-ε.c.x), where x is the distance in the cell from the light entry window, c is the absorber concentration and ε is the molecular extinction coefficient. It is expressed by:

$$Ia(x) = \varepsilon.c.F(0).\exp(-\varepsilon.c.x) = F(0).\exp(-x/\mu)/\mu$$

where F(0) is the photon flux density at the considered wavelength.

Oxygen is an inhibitor/quencher of polymerization, and as long as it is not consumed, the medium remains fluid: thus, in the presence of light, it is possible to create a "real" polymer layer because the resin reacts and oxygen is consumed at the same time. Additionally, these layers adhere to each other, enabling the creation, layer by layer, of a 3D object (André 2017a). However, in this case, the definition of the problem (defined by the expression additive manufacturing) is inherently poorly known and not yet structured. It is a challenge in its own right, since the process is new (apart from the Battelle patents, there is little or no theoretical and practical knowledge to guide the design work, at least in the inventor's mind).

On this basis, it is not possible to adequately reach all points in space to build an object, which imposes working on the surface (2D1/2), layer by layer. This is how stereo-lithography came into being (André 2017a)! Although the (divergent) idea is simple, its simplicity is a far cry from research in design-science (RDS), which are incremental and organized around major stages (problem identification, artifact design, artifact evaluation). They recognize the importance of relying on available theoretical knowledge to build the artifact that responds to the initial problem (Hevner et al. 2004; Peffers et al. 2007; Sein et al. 2011; Pascal and Rouby 2017).

Today, additive manufacturing is made up of seven standardized process families, according to the French standard NF: E 67-001 and the international standard ISO 17296-2:2014E. These are the following processes (MADEELI 2017):

- vat photopolymerization (stereo-lithography);
- material jetting;
- binder jetting (powder bed);
- powder bed fusion;
- material extrusion;
- Direct energy deposition (DED);
- sheet lamination.

More recent methods range from the production of nanometric to decametric objects. Some are true 3D printing using for example, two-photon stereo-lithography (Doualle et al. 2021).

However, the clarification of the invention's presentation is crucial, because it needs to be understood if it is to be "desirable" and attractive to future users. In the digital and television age, an essential contribution has been the visual presentation of the innovation based on POC and tangible results (see Figure 2.3). It is an integral part of the designer's experience, which is in essence little-known by target users and still poorly structured on the design side. At this stage, this conceptualization can be aimed at an initial understanding of the host sociotechnical system and its role as a lever for further development (from the scientific and technological points of view), including people likely to be interested in the operation, some particularly illustrative examples of the interests of the invention, clear messages (for example, the idea of a 3D photocopier), without mentioning difficulties, and problems considered secondary (Carlsson 2007, 2010).

So, in this perspective of promoting their invention, the researcher steps out of their role as "scientific and technological" inventor, because they are obliged to go further by enriching their activities with a real demonstration that must integrate, at least in part, future users' knowledge of their daily work practices and the organizational context in which they might operate (Pascal 2012). It needs to communicate.

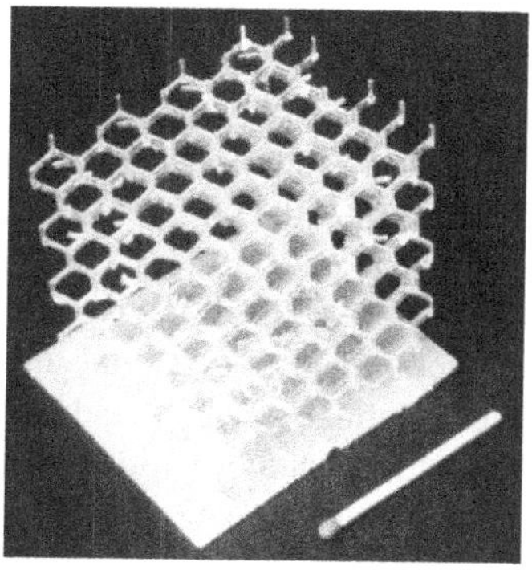

Figure 2.3. *Example of a 3D object produced in the late 1980s*

To clarify the objectives and corresponding expectations of users, it is useful to go a little further than "images" and "videos" to convince them:

– specifying the context of responsible material interest, material gain, user image;

– the work context, where the user is the "authority", with social and even material interactions outside of work (networking, sharing, etc.);

– the disruption of the organizational context that governs and regulates new localized and networked work practices in the uses of new technology.

NOTE.– 3D printing.

If, in 1984, we had access to coherent sources suitable for the resumption of Battelle's work, it would have been possible to produce objects in polymers, preferably acrylics, from transparent resins (see the question of optical thickness). These objects, new from a scientific point of view, might well have been noticed, but no economic development would have been envisaged (for reasons of cost in particular, and the impossibility of producing objects in "good material"). The "frugal" approach using a low-cost, single-photon absorption process led to the concept of additivity and, ultimately, to other additive manufacturing technologies, just as disruptive as stereo-lithography.

If there is one thing engineers know well, it is that they will have to modify or redesign their prototype during production. Additive manufacturing gives them the freedom to redesign a part and innovate on the spot, without major delays or financial consequences. This freedom has many advantages: shorter production schedules, improved product quality, model diversification and, ultimately, increased revenues (Tranchard and Rojas 2015). It is important to arrive on the market with a perfectly validated quality product. Decisions taken at the design stage have a profound impact on total manufacturing costs, and therefore on the overall profitability of the product. It is accepted that 75% of total development and industrialization costs are determined at the very start of design (AFPR 2015).

Rapid object materialization, from a rough but rigorous digital definition, is now possible thanks to the rapid development of computer-aided design (CAD) software. This is why rapid prototyping techniques, which lead almost in real time to models that can be used to validate the design and/or validate functional tests, are making a vital contribution, made easier thanks to additive manufacturing. Thanks to the existence of new and useful performances, the industrial world is ready to accept 3D technologies right from the start.

With a continuous digital chain, rapid prototyping, through its associated methods and resources, enables all those involved in the product to:

– test alternative solutions as early as possible;

– validate the technical and/or technological solutions chosen for the future product: appearance, ergonomics, processes, etc.;

– work in parallel and in a coordinated way in the context of concurrent and simultaneous engineering to optimize the product.

These different contexts need to be described simultaneously and in relation to each other (Go and Carroll 2004; Stahl 2011; Carroll 2016; Pascal and Rouby 2017). Today, 3D printing is often presented as an essential vector of a possible "industrial revolution". Its impact is significant (around €30 billion/year, with a growth rate of around 20%/year) without being major, but we need to counter some recurring misconceptions about it. This is particularly the case when it comes to the freedom of design and functionality of objects. Even if additive manufacturing processes allow for the exploration of new possibilities, this finds limits linked to different types of technical, educational, economic and/or cultural difficulties, which have already been mentioned in André (2017a, 2017b, 2017c).

This fairly recent generic technology (apporx. 40 years old), apart from its material-saving its performance, as shown schematically in Figure 2.4 (Conner et al. 2014; DIRECCTE 2014; Mizeret 2015; Pinkerton 2016), today enables the production of small runs of complex parts inaccessible to conventional methods for the same time and cost (see also Mellor et al. 2014; Weller et al. 2015). What is more, at the end of the IT chain, this process, without (too much) direct human intervention, authorizes a specific relationship with the object (a relationship of the same type as that between a citizen and their information and communication systems). In addition, 3D printing is a sustainable alternative to mass production, since it reduces the waste associated with the loss of material during manufacturing (known as subtractive manufacturing). This is particularly true for metal objects, where around 80–90% of the original metal is unused.

From an operational point of view, the scenario is essentially intended to describe the day-to-day progress of the user's work and tasks (it can then be referred to as a task scenario). In particular, it aims to specify information flows, the nature of the information expected and/or provided, as well as its content and/or structure, the user's direct partners, and all of the material resources used. From this perspective, the theory of distributed cognition is one of the most widely used theories for constructing scenarios (Hollan et al. 2000; Carroll 2003; Stahl 2011). Here, tool/user interaction is seen as rooted in a local work context made up of individuals interacting with each other, and with tools, in a distributed cognitive environment. Tool use is seen as articulated to that of other devices, and thus depends on its ability to integrate with the repertoire of pre-existing tools. It also depends on its ability to contain and structure information that makes sense to its users, given the specific tasks they have to perform, their own knowledge system, and the knowledge available in other tools.

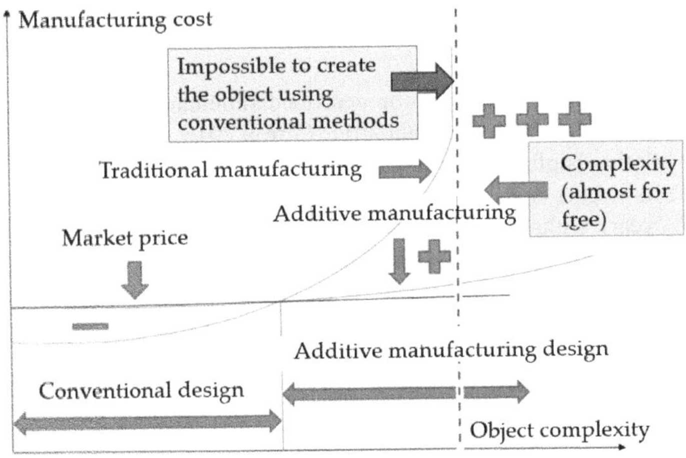

Figure 2.4. *Advantages and limitations of 3D printing in relation to object design of an object (the – sign represents the advantageous area for traditional design, the + for additive manufacturing design which becomes +++ when the object cannot be produced using conventional methods)*

Cornu (2008a, 2009) wrote: Apart from its undeniable progress, does technology serve to fuel dreams to make applications more acceptable to the greatest number, or does it give a few science-fiction fans the legitimacy to influence research programs? Whatever the case, a "convergence of interests" has enabled us to rally around this mobilizing program.

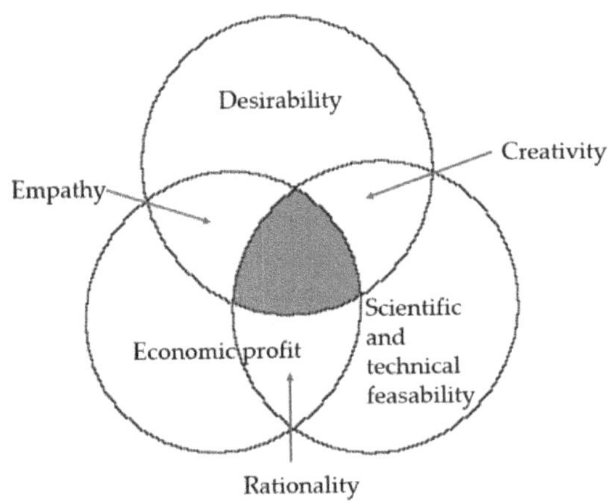

Figure 2.5. *The foundations of successful innovation*

3D technologies fit right in! Additive manufacturing is the result of a combination of scientific, technological, economic and social aspects, as shown in Figure 2.5, for successful innovation (Buttin and Saffré 2016). It is the articulation and integration between these partially disjointed components that where we can judge whether technologies are "coming of age". The odds are looking good.

2.2.2. *"In-house" activities*

> Additive manufacturing [...] is seen as a revolution, but it has not taken place, we are on a step-by-step evolution, and that is what is important. We must continue to work on it through the evolution of materials and the evolution of machines (Viénot 2022).

> What can defeat the system is not positive alternatives, but singularities. And singularities are neither good nor negative. [...] They can therefore be the best or the worst. They cannot therefore be federated into an overall historical action. They thwart all single, dominant thought, but they are not a single counter-thought – they invent their own game and their own rules of the game (Baudrillard 2002).

> The intentions we form in our day-to-day lives take on a greater or lesser degree of importance, subjectively assigned according to our values, desires, goals and the more or less dramatic consequences an oversight can have on our lives. This level of importance modulates our motivation and the strategies we put in place, which has a major impact on the likelihood of carrying out the intention (Lecouvey et al. 2015).

Communities with a shared disciplinary passion are normally innovative, but sometimes tend to close in on themselves, in their "keeps". This is reflected in work that respects the paradigm "of the moment": academic societies, specialized scientific journals, etc. The good fortune for additive manufacturing, after a personal POC, from a scientific point of view, was to be able (to be obliged to) develop an interdisciplinary, heuristic, temporary and open space in research, authorizing a more collective trial and error operation of the creation and exploitation of new concepts; this is becausewe quickly realized that the demand was there. Thus, by association with scientific disciplines, on subjects of public utility, we have been able to bring out discontinuities inducing scientific recompositions and adaptations. These are the components that must constitute the specificity of the development of the field of 3D printing, aiming, through new approaches, at other links between humans and artifacts, between workers/users and production systems/artifacts.

This is why we need to account for the way in which the objects studied by scientists and technicians return to the social world, a question largely addressed in other chapters. It is a central question that strongly links research to the economic, social and political, and puts additive manufacturing in direct and responsible contact with society. Today, it is all about fostering the need for greater trust between all of the partners involved in this disciplinary hybridization. So, proposals have been made (André 2017b) for researchers to consider innovations more as ends and less as means (especially for funding).

To this end, all useful scientific and technical avenues must be exploited to satisfy an intentional research objective on material or non-material "objects/ devices":

> Either the researcher treats the object in a partial way, ignoring or concealing all the links that inevitably exist between this partial view and the overall biases that he cannot fail to have on the object. Or else, he puts his global parties at the heart of scientific questioning: not as interesting elements to watch, but as the motors of a dynamic which then takes the inevitable desire for unity as a stepping stone towards this which becomes a requirement of unity (Alvarez-Pereyre 2003).

This desire to get closer to reality, with all its difficulties (Nicolescu 1996), must define a common cultural background, that of modeling and integration (systemic approach), combined with experimentation, as well as an indispensable openness to other scientific and technical fields; this is by "borrowing" their concepts, methodologies (principles of constant learning and cross-fertilization), their instruments, by "bringing in" subjects for study, etc. All disciplines are involved from the "hard" sciences to the humanities and social sciences.

Therefore, to achieve the ambitious goal of bringing together the scientific and technological forces behind the development of 3D printing, we need to be clear about our subsidiarity (a shared cultural heritage), while at the same time being clear about the relational approaches we need to consider in order to develop the desired interdisciplinary openings. Indeed, bringing disciplines closer together can encourage innovation at the interface, through unexpected connections, project mode, exploration of the same instrument for different purposes, etc. (Vinck 2007). This is indeed what has been observed with, from the concept of additivity, the emergence of other creative techniques of additive manufacturing, or even 3D printing, recalled in Table 2.1. As these are clean breaks, not requiring (too) complex approaches at the start of the invention, it was not necessary, as for stereo-lithography, to take into consideration the difficulties of interdisciplinarity and therefore efforts to support actions with decision-makers so that this creative discipline develops. In fact, it is rather a question of multidisciplinarity.

In reality, there was a need to do some engineering work, a little less rustic than this presentation based on masks might lead you to believe, by exploiting everything that was accessible in our "toolbox". Playing on the process, the choice of light sources and the optimized control of the light beams, the materials and their formulations, the choice of photochemical initiators, the computerized control of the galvanometric mirrors, which precisely position the light beam on the surface, then control digitally, the production of a piece (with its supports) was an exercise in intellectual acrobatics or "scholarly" DIY which went somewhat beyond "classic" interdisciplinarity or multidisciplinarity.

First, validation of the 3D process was rapid and achievable by a single person, since it involved taking existing elements and assembling them in a new way, and second (perhaps fortunately), because disciplinary support (human and financial) was provided sparingly. It may have been a matter of immediacy, emergency management and, at least at the outset, a very small-scale affair, but it was "learning" research, and practically free of charge. Then, thanks to partially "diverted" funding, it was possible to propose technologies that went beyond POC, with UV, visible or IR lasers, for the production of objects. More and more complex parts began to emerge from our machines to our great satisfaction (it was magic!), and more than 20 patents were registered.

But with success, the research component took up the subject, bringing various incremental improvements (see Figure 2.6; adopted from Chenavaz (2012)): the time effects of incremental innovations are reduced, production quality increases, machine prices and running costs decrease, while the attractiveness of 3D processes increases.

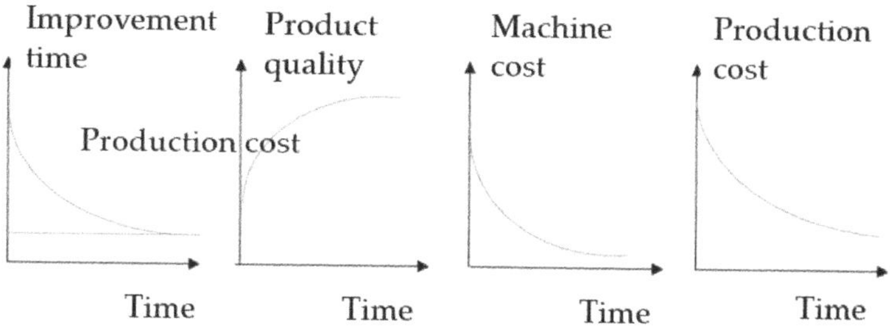

Figure 2.6. *Temporal effects on process improvement*

NOTE.– Epistemology.

How can epistemology be useful to the engineer as a scientific and technical player? It is still the prevailing belief that engineers are more likely to be users of scientific knowledge and technical tools than creators of them, even if they do sometimes design new tools. In fact, engineers are involved in action, most often technical action. In the course of their work, they have to draw on knowledge and know-how from a wide range of sciences and techniques, and the question of interfaces between them arises. These interfaces are generated by scientific research, which is generally built on a precise definition of its subject; physics, chemistry and biology, for example, have different scientific issues. But for the engineer, knowledge of sciences and techniques cannot be separated from their use; there is therefore a panoply of engineering sciences and techniques. This leads to a praxeology (philosophy of action in the methodological sense), a reflection on scientific-technical decision-making, and ethics (philosophy of action with reference to values: good or just action) (Dubois and Brault 2021).

"Doing science" is not an absolute or a neutral effect of a neutral decision. Every scientist, without necessarily seeing this as a renunciation of objectivity, questions nature in coherence with their conceptual scheme, a vast latticework of more or less stabilized and semantically qualified concepts. "Engaging in science" is therefore a twofold act: (i) ontological engagement: the choice of a conceptual exploration of reality; (ii) axiological commitment: the reasons for choosing science as the exclusive means of attaining a knowledge of the real – by virtue of the decree of reason enjoining us to consider the course and methods of science as the most adequate instruments for the project of making account of the world – and, therefore, as a major goal of culture (Silberstein 2013).

2.3. Reaching out to society

> Cognitive intelligence has become the sole criterion for meritocratic selection; the head has taken over from the Hand and the Heart, and shaped society to suit its [own] interests (Goodhart 2020).

> The confusion [between rites and techniques] is all the easier because the traditional character of magic is found in the arts and in industry. The series of gestures of the craftsman is as uniformly regulated as the magician's series of gestures. [...] In techniques, the effect is conceived as mechanically produced. We know that it results directly from the coordination of gestures, machines and physical agents (Mauss 1950).

> Official STI policy that stresses the importance of making science intelligible and how science is a common good for all citizens. However, these mostly seem to perpetuate the 'science for society model' and the 'deficit model'. France's formal administrative style to offer clear prescriptions on how to operationalize these responses, but at the same time risks precluding responsiveness and creativity (Wittrock et al. 2021).

The innovation is thus introduced by the inventor in a context that Schumpeter (1942) calls "destructive competition". In practice, this gives rise to the possibility of industrial competition, associated for a time with a series of duels between the accepted existing and a new invention. On this subject, Tarde (1902) writes: "The only utility [...] of opposition is to provoke a tension of antagonistic forces capable of arousing inventive genius, the military invention which, by giving victory to one side, momentarily puts an end to war, the industrial invention which, adopted or monopolized by one of industry's rivals, ensures its triumph, and momentarily puts an end to competition...". Alerting users to the benefits of additive manufacturing is one thing; thinking that all manufacturing will go through this process is another! A dynamic balance is being found.

To be convincing, scientific knowledge is essential to promote the expansion of inventions in industry. Therefore, for Hohberger (2016), understanding the influence of stabilized scientific knowledge on the diffusion of innovation processes is essential to understand, predict and envision future patterns of invention (see Tables 2.1 and 2.2 for examples of alterative processes and evolutions toward the satisfaction of "good material" needs). However, the inventor still needs industrial and/or financial support to convince other components of the socioeconomic world.

NOTE.– Publish to be known!

Sorenson and Fleming (2004) identify the importance of publications by comparing citation patterns from peer-reviewed documents with those from non-scientific sources (commercial documents). They consider that publication, whatever its nature, contributes to accelerating the rate of technological innovation. Insofar as peer-reviewed publications capture other mechanisms of a scientific nature, such as differences in the process of solving problems, they suggest that this contribution to the development of an idea remains low, relative to more dynamic and convincing communication. This is an interesting consideration, given that, thanks to television, the 3D process became known, whereas the first scientific publication on the subject was rejected because the process was unknown. Where are the conservatisms?

Process	Matter or materials	Advantage(s)	Disadvantage(s)
Stereolithography (SLA) (1984)	Acrylic resins, etc. Possibility of using various additives particles (ceramic or metallic)	Manufacturing speed Surface state	Implementation of media for manufacturing Post-treatment
Material projection (1987); process invented by DTM corps – PJET Poly-Jet	Thermoplastics, photo- polymers, waxes, etc.	Manufacturing speed No need for scaffolds	Machine cost Deformations
Binder Jetting (BJ); Stratasys founded by Scott Crump	Numerous mineral and/or organic materials,	Manufacturing speed No need for scaffolds	Machine cost Precision and roughness Lower deformations than other technologies
Selective laser Sintering (SLS) (patented by Carl Deckard in 1988)	Polymers, Metals, Ceramics	No need for scaffolds Materials similar to those used traditionally	Porosity of parts Heat treatment necessary Deformations Significant roughness
DMLS (Direct Metal Laser Sintering) (1995)	Metals sintered directly	No need for scaffolds for manufacturing	Need for scaffolds for manufacturing Recycling
BJ Color by ZCorporation	Colored powder materials like in traditional printing	First colored objects created "simply"	Moderate mechanical strength
Carbon 3D (2014)	Fluid acrylic resin	Manufacturing speed Clever Process	Not all forms are accessible

Table 2.1. *Other industrially important 3D processes (source: Protolabs 2020; Philippe 2021; Rodriguez 2021)*

Process	SLA	SLS	DMLS	FDM	BJ	PJET
Quantity						
Complexity						
Surface finish						
Choice of materials						
Life-time						
Color						
Tolerance						
Manufacturing speed						
Price (unique parts)						
Price (series)						

Table 2.2. *Potential of different 3D technologies: red: rather negative; light blue: possible; green: positive*

Technology transfer from academic research to industry depends on various concepts of innovation processes. The so-called "linear" model of innovation (Bush 1945) assumes a frictionless, unidirectional transfer without the active participation. For Palmberg et al. (2010), this simple vision has been made more complex by the interweaving of interactive models. This two-way street allows researchers and companies to interact intensively, and quite frequently relies on research external to the company (Kline and Rosenberg 1986; Cohen et al. 2003). But to accept this consideration is to accept that the origin of the idea has passed from the researcher to the company! This was not the case with 3D printing! Future engineers are trained in scientific knowledge, but not in technical problem-solving.

The original technical fact is resistant to snippets of comparisons, even if we assume "all other things being equal", because the rigidity of its new anchorage, in the material world, may discourage inappropriate extensions; it is just new! In the case of additive manufacturing, this disruptive technology has introduced new values to the market, relative to what was available through established technologies, products or services (Christensen 1997). This invention offered different functionalities from the initial offering: 3D printing thus offered original attributes, attractive to a specialized part of the market (Christensen 1997; Christensen and Raynor 2003). For example, with 3D printing, companies have been able to adopt just-in-time manufacturing to remain agile, customize their products and stay in tune with changing market demands.

It is now well-known that the industry must innovate while being subject to various constraints and, to do so, improve the development cycle of products and/or processes from the idea to the product launch, with respect for the following principle: an ever lower price, a varied choice, availability and satisfactory quality. To respect this principle, several strategies have emerged: offshoring, manufacturing of intermediate products, technology transfer, automation of production, new manufacturing systems, development of logistics, additive manufacturing, etc. With regards to this last area, what has been shown is the possibility of strategies that correspond to the development of innovative manufacturing systems that can meet these requirements (development of productivity, then flexibility, industrial management and automation, industry 4.0, etc.). Additive manufacturing demonstrates great reactivity to responding quickly to the design and then to the appearance of a new product (design evolution) and/or a new demand (volume fluctuation), which keeps it efficient. Additive manufacturing is a new manufacturing system capable of ensuring high productivity and high flexibility.

Additive manufacturing makes it possible to produce physical models, regardless of their complexity or almost, which, it should be remembered, can sometimes be impossible to produce in one go by subtractive manufacturing. Apart from the elements that have already been presented, there are other advantages to using these processes:

– rapid detection of design errors;

– testing alternative solutions on part components;

– validation of industrial feasibility;

– optimization of object shapes;

– facilitation of product modifications;

– refinement of the object's operational characteristics;

– mechanical and thermal testing, etc.;

– provision of a support object to avoid potential conflicts.

Aladad (2009) points out that flexible manufacturing systems (FMS) or flexible workshops (FW), developed in the 1980s, are expensive and often under-utilized compared with dedicated transfer lines (DTL) or dedicated manufacturing systems (DMS), which are highly productive but not very scalable. So-called reconfigurable manufacturing systems (RMS) have been proposed to meet these two conflicting constraints, but they do not offer the same advantages as additive manufacturing (particularly in terms of customization). On the other hand, with these 3D technologies, there is little waste. Their environment is decentralized, and the use of 3D printing can go rapidly from prototyping to finished product (André 2020). This environmental aspect is now a significant advantage (see Figure 2.7; taken from Despeisse et al. (2017)).

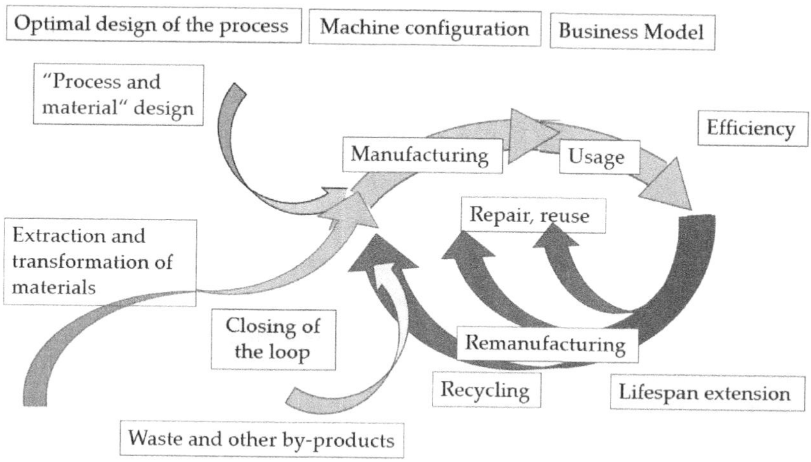

Figure 2.7. *Positive environmental aspects of additive manufacturing*

2.4. Consequences

> 3D Systems has taken advantage difficulties of Phenix Systems – 50 employees, €5 million in sales, bought out for €15 million, a pittance! – to acquire the French company's technology. The company created in 2002 has not found the refinancing to continue its expansion (Battini 2015).

> Energy is a kind of fantastic projection; projection that feeds all the industrial and technological dreams of modernity, it is also what inflects the conception of Man in the direction of a dynamic of will. However, we know, by the analysis of the phenomena of turbulence, chaos and catastrophe in the most recent physics, that any flow, any linear process, when it is accelerated, takes a curvature strange which is that of the catastrophe (Baudrillard 1990).

> Because confidence results in motivation, which is so necessary when it comes to making efforts, investing time and resources, and remaining persistent in achieving goals (Goetzmann 2021).

We discussed innovation through active filters in Chapter 1 of this volume, calling them "death valleys". On the French side, as long as the research team remained in control of the operation, it was able to enjoy the freedom (and funding) needed to achieve the POCs that led, through good communication, to effective recognition (Prix national de la créativité in 1990).

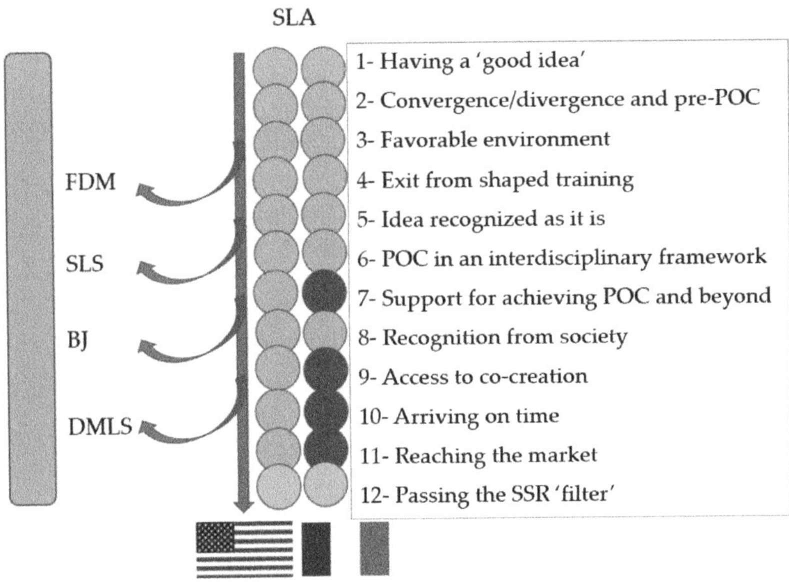

Figure 2.8. *"Valleys of death" and 3D printing (specific case of stereo-lithography)*

The basic patent (André et al. 1984) belonged to Alcatel (for CILAS, Compagnie industrielle des lasers), which saw no point in maintaining this patent and, without talking to the CNRS, which was to be the co-owner, or to the inventors, put it in the public domain. This was not the case in the United States with the creation of 3D Systems (see Figure 2.8). Several lessons can be drawn from this blockage:

– The relationship of trust between partners does not have to be considered as soon as there are financial operations associated with a project.

– Decision-makers do not have access to up-to-date knowledge, prefer the short term and, above all, avoid all risk-taking. The pioneering spirit is to be found on the other side of the Atlantic.

– By losing (at least partial) ownership of the first patent on additive manufacturing, it was not possible to create a national company, as the French only had to buy American equipment.

– To go further in terms of R&D, we had to seek out contracts that led to new patents, resulting in a series of protections, but with increasingly numerous owners or co-owners.

This is a good lesson! According to the NCSE (2021), the root causes of failure can be defined when a product or process fails to reach the market:

– There were either no requests or very few: FALSE.

– The product or business process did not perform as expected: FALSE.

– The adjacent business inputs needed for customers to use it successfully were not sufficiently available: FALSE.

– Regulatory restrictions on its use have limited its economic value: FALSE.

– A product or business process has been developed, brought to market and aroused positive interest. However, the innovator has not been able to secure a solid position for the innovation in the market, for example, through intellectual property protection: FALSE.

It would be interesting to examine in advance the real causes of such a failure (tens of billions of euros per year!).

2.5. 4D printing

> Big Bang disruptions often come out of nowhere, from people who are not your traditional competitors. They are often developed by inventors who are simply experimenting at low cost with existing technologies to see what new products they can invent. Once launched, these innovations don't follow traditional strategic paths or normal patterns of market adoption. This makes them incredibly difficult to combat (Downes and Nunes 2013)[1].
>
> Thus, the completion of the interconnection of systems in globalization and the effects and feedbacks of human action on them mean that the practices of thought, those of science and those of action, can no longer define a 'temporality' object that would be their own or that could be abstracted from its 'incarnations' for study? (Theys and Cornu 2019).
>
> Discoveries are made by people who are curious about what's new and unprecedented, and dissatisfied with the status quo. They are curious people who ask old questions in new ways and ask new questions (Bonasso 2007)[2].

1 Author's translation from the French version of the text.

2 Author's translation from the French version of the text.

Kelly et al. (2021) have developed an original method for detecting an important patent or a revolutionary innovation that stands out from previous industrial protections (indicating a new idea). However, this type of signal, important for the 4D community (designers and users), could not be found because the number of patents on the theme is very low and with a modest generic aspect. A specific set of skills is required, such as the ability to "translate" and build bridges between basic research and R&D. It is on this basis that Assmus and Haeussler (2017) defined the ability to transform discoveries into products. From a scientific point of view, novelty must be an integral part of the work of the researcher or creator, but in the presence of other constraints, it must remain within an "acceptable" framework. This is what happens for 4D printing, which will be discussed below.

After Tibbits' radical paper in 2013 (Tibbits 2013, 2016; Tibbits and Thomas 2013; Thomas et al. 2014), displaying an open and original research path consisting of associating with 3D printing adaptive materials that can see their form and/or functions evolve under the influence of an energy, a framework was created. It introduced historically established systems of meaning to support the idea. Indeed, understanding and interpreting a new event involves invoking pre-existing symbols, codes and narratives, to help us make sense of the new idea (Mast et al. 2013), even if, as with 3D printing, it exploits, in an original way, stabilized knowledge. Most of our activities continue to be predictable and determined by our past knowledge and experience. We generally forget that our familiar ways of conceptualizing reality are the direct result of our education and training experiences (Bonasso 2007).

As Roco and Bainbridge (2013) point out, science is made up of a collection of more or less airtight boxes, and the work of integration makes it possible to assemble disjointed knowledge. Since initial training creates formatting, it is sometimes difficult to escape this conservatism. One of the influential conceptual frameworks for thinking about research goals has been proposed by Stokes (1997), who advocates research in what he calls Pasteur's quadrant to answer affirmatively to the following two questions: "Are considerations of the practical usefulness of the results crucial?" and "Is research a quest for fundamental understanding?". More visionary research may also meet these two criteria, but adds two more questions: "Does the research lead to emerging uses beyond known applications?" and "Is the work transformative in the sense that entirely new ideas are explored and invented?". Figure 2.9, considering the domain "Basic research inspired by a vision", extends the four existing domains of the Stokes diagram to a new element devoted to research oriented toward new applications ("new use"), beyond the Pasteur quadrant which has been defined for predetermined applications ("known use").

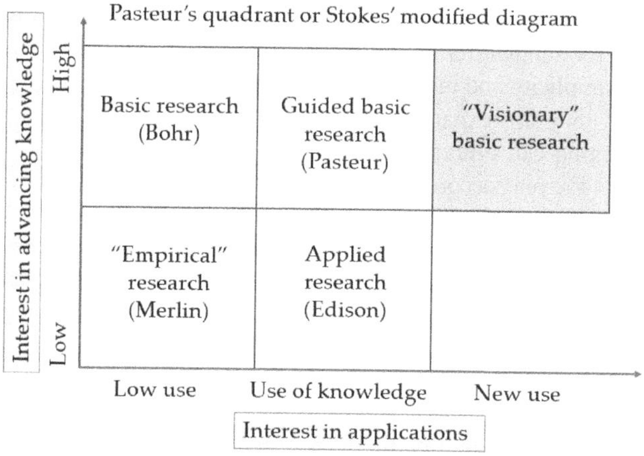

Figure 2.9. *Pasteur's quadrant: three basic research approaches (pure, use-inspired and vision-inspired) are proposed for implementation according to the respective phase of the convergence-divergence process of knowledge and technology: creation phase, integration/merger phase and divergence phases, respectively*

The invention process applied to 4D printing covers all efforts to create new ideas and make them work for their application. This means considering, as soon as possible, the stages of commercial development, application and transfer, including the redirection of ideas or inventions toward specific profitable objectives, the evaluation of these objectives, the downstream transfer of research and/or development results, and the large-scale use, distribution and diffusion of technological results.

It may also be useful to take into account "market pull" as a demand for innovative projects, or even "technology push"; that is, projects to improve the technical state of the art without taking into account the specific commercial benefits that might result, etc. The overall management of the technological thus includes the organization and orientation of human and capital resources toward efficiency:

– creating new knowledge;

– generating technical ideas to create or improve products, manufacturing processes and services and services;

– developing these ideas into working prototypes;

– transferring them to manufacturing, distribution and use.

For BFF (2022), “the automotive, aerospace, electronics, pharmaceutical, construction and food industries are the main fields where 3D offers real added value. In construction, simplicity and innovation are its main assets”. There are many uses for 3D printing, and they are expanding rapidly. At this point, it is worth considering whether “4D printing can offer the same advantages, or even more...” . In a rapidly evolving field (+44%/year according to Demoly and André (2022)), it is useful to try to understand how the most creative researchers work and think about invention and/or innovation, and how to encourage their 4D-oriented creative productivity. It is also useful to try to understand how the organizations that fund them integrate and exploit their dedicated 4D creative ideas. Finally, it is relevant to understand how market mechanisms can manage/frame/support this creativity and how research stimulation policies need to be adapted (Burger-Helmchen 2013; Ogilvie 2019).

Nevertheless, even if there is time pressure for innovation in 4D, it may be important to distinguish when and where the science needs freedom, space and time to engage with uncertainty.

> Conditions must be created to enable and encourage serendipitous discovery. When and where do we accept that it is not possible to state in advance what research will find or which applications will ultimately emerge from seemingly useless knowledge? (Nowotny 2015).

NOTE.– Fab-Labs.

There have been several waves of evolution in Fab-Labs (Damoah and Botchie 2020):

– sustenance and the DIY spirit (Fox 2019): DIYs are made for personal use (Jeacle 2017);

– industrialization: commercialization of DIY activities (Fox 2019);

– innovation: technology and entrepreneurship: the use of DIY in original proposals for science, innovation, technology and entrepreneurial activities (Landrain et al. 2013; Fox 2019; Sarpong et al. 2020).

In this last area, it will be interesting to see whether 4D printing has been able to take advantage of these flexible organizations to innovate.

> The spread of do-it-yourself activities in the United States, especially over the last ten years, is considerable [...]. For those who manifest a preponderant ‘dada’, whatever it may be (and the list is practically infinite, from photography to scale models, from tapestry to philately [...]), there can be a natural restructuring of the personality. Some observers even predict, in the society of the future of the future, the

> blossoming of a new Artisan Man, dedicated to the patient and creative shaping of materials using manual tools, a new *Homo Faber* resurrected by leisure activities (Friedmann 1963).

2.5.1. *Reminders about 4D printing*

> Innovation project: made up of interaction, decompartmentalization, information flow, adaptation and flexibility. The collective player must be able to react to all fluctuations and seize all opportunities. To innovate, therefore, we need to avoid rigid, mechanical models, overly precise definitions of tasks and roles, and overly restrictive programs (Akrich et al. 1988).

> A single person can promote an idea, but he quickly surrounds himself with a few individuals who share the same convictions and aspirations. The innovation process is necessarily collective (Gaglio 2011).

> Creation is inseparable from a willingness to let go of oneself (Caillat 2014).

> In the complex sciences, we have knowledge of everything. Ideally, we should no longer be specialized or mono-disciplinary. The field of complex systems can be seen as a response to the over-specialization of research. But as a rule, the researcher remains a specialized beast. (Valke 2021).

New application niches are emerging, combined with the stabilization and robustness of the 3D technology, which serves as an increasingly robust foundation. With the use of more "intelligent" materials and the development of alternative processes, it has been possible to push 3D technologies beyond their fields of excellence to explore new application niches, such as those seeking to modify the shape and/or functionality of 3D objects to create new motors, actuators and/or sensors in a single operation, contribute to the development of flexible robotics, etc. The biomimetic approach (like the pinecone that opens in the absence of moisture, so that the seeds reach the ground at the "right" moment) serves as a good pictorial representation for presenting and exploiting the concept (Graeff 2020).

With 4D printing, some may have thought that we could free ourselves from the determinisms imposed by 3D machines by using physicochemical modes of material transformation to play on spontaneous or stimulated self-organization processes (see the particular case of bio-printing, a form of 4D printing applied to living matter). While POCs can highlight a very broad field of possibilities, in reality, and for a variety of reasons (scientific conservatism, satisfaction of need, use of commercial 3D machines, etc.), we are still a long way from *The Terminator*, where matter can be controlled by the mind. Moreover, it is unlikely that a given shape corresponding to a precise set of instructions can be achieved from specific spontaneous, or even controlled, perturbations in space and time. While it is possible to direct a three-dimensional system, at least in part, toward a desired shape, by introducing an apparent determinism sometimes controlled by the engineer, based on certain laws of physics, chemistry and biology, there are still many obstacles to the development of 4D printing. However, by modifying the environment, it is in principle possible to change the shape of the objects created in one or more potential fields (mechanical, electromagnetic, photonic, chemical, biochemical, etc.). This is what defines the basic idea of 4D printing, which brings an essential element – time – to additive manufacturing, already covering the three space parameters (André 2017a, 2017b, 2017c; Demoly and André 2022). Figure 2.10 shows the foundations of this recent technology.

For example, 4D printing uses products or materials that are printable and participate in the realization of a 3D object (see, for example, Listek 2019; Schwartz and Boydston 2019; Shen et al. 2019; Ahmed et al. 2020; Fang et al. 2020; Goo et al. 2020; Jeong et al. 2020; Wang et al. 2020; Zolfagharian et al. 2020). Programmable materials/objects could build themselves, making assembly plants and cumbersome installation procedures super-fluid. Robotization, at the heart of productivity gains in the 20th century, could thus be integrated into the products themselves. According to O'Neal (2020), 4D printing has great potential for personalizing devices and structures. For the moment, known applications are more the subject of projects than actual implementations (see Chapter 3).

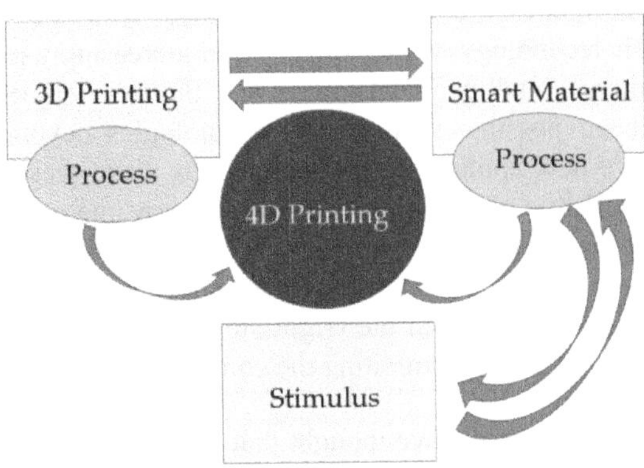

Figure 2.10. *Process coupling – materials in 4D printing*

Current experimental results on 4D printing are based on spectacular visual experiences (POCs) which, despite certain claims, are neither thought of nor experienced as likely to be generally applicable. These experiments are used for numerous theoretical developments and modeling (Demoly and André 2022). The emphasis is rather on singularities and not on applicative processes (paradigms). This singular aspect of 3D printing will be discussed further in Chapter 3. Convergence indicates a decisive activity, the sorting out for (correct) diagnosis – it is an inverse problem that needs to be tackled, starting with the goal and working backwards to the phenomena to be implemented. In fact, to be reductive, scientific work is too marked by the sign of a certain continuity (as a result of spectacular results), of the unfulfilled promise (promise of applications), in short, by forms of discrepancy between the felt need and 4D-oriented scientific knowledge and applications.

With the same or different materials having variable responses in the presence of energetic stimuli of various origins, the shapes of structures can evolve spatially and over time. Figure 2.11 shows some typical effects sought in 4D printing (Yao et al. 2018; Demoly and André 2022) with a spectacular example, presented in Figure 2.12 (Ge et al. 2016).

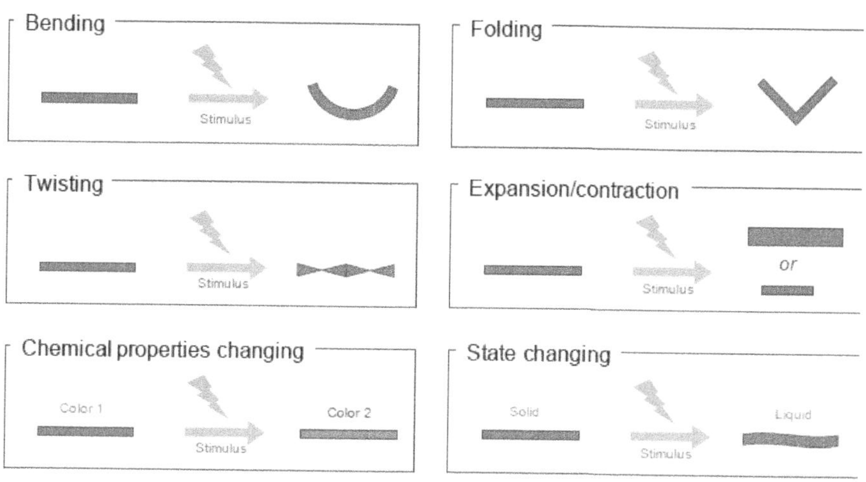

Figure 2.11. *Possible deformations of pure and/or associated materials that can be used in 4D printing*

Figure 2.12. *Spectacular example of a 4D object made with heat-sensitive polymers, validation in additive manufacturing (source: Ge et al. (2016))*

2.5.1.1. *The risk of a dead in the water technology*

Once a discipline, methodology or a process has been established, groups of players with shared visions are formed (learned societies, networks of players, etc.). From a historical point of view, the example of Thomas Edison can be recalled, imposing a point of view on the use of direct current (Corazza 2015). This situation, which may have met societal needs, was challenged with difficulty by Tesla, imposing the use of alternating current. But it was necessary to go beyond corporatism, habits and a lack of willingness to take risks, which is always a difficult process, especially if there are no deep-rooted motivations or credible alternative solutions. To illustrate this point, it was possible to produce Figure 2.13 by reading the titles (and numerous abstracts) of the various articles associated with the terms 4D printing, taken from the CNRS database. Several keywords were selected, corresponding to (1) organic materials and their stimuli, (2) inorganic materials and their stimuli, (3) synthesis articles, (4) modeling, (5) miscellaneous, (6) applications and (7) processes (including 4D origami).

What is important to highlight is the very high degree of unity in the work focused on the stimulation of pure organic materials (polymers) and charged materials (in particular nanoparticles), accounting for 63% of published activity. But with 16% of publications in synthesis (for an emerging field where the number of publications is in the hundreds), this percentage rises to 79% of the total. And when we remove miscellaneous documents and applications, the figure is even higher.

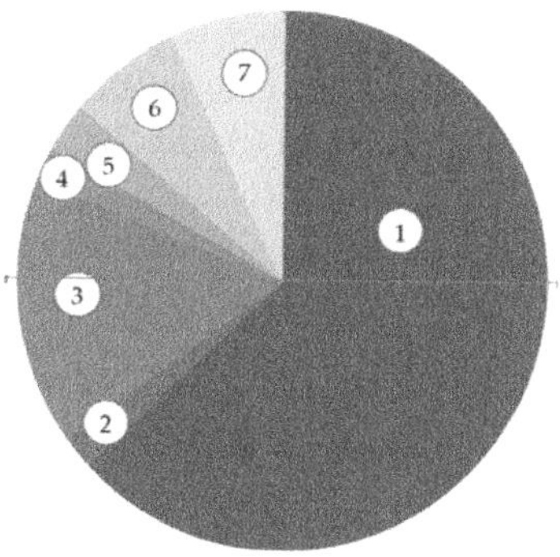

Figure 2.13. *Topics covered in 4D printing publications (1: 63%; 2: 2%; 3: 16%; 4: 2%; 5: 3%; 6: 7%; 7: 7%)*

From an experimental (and partly theoretical) point of view, apart from the focusing of research on targeted aspects, other problems have been noted (André 2017a, 2017b, 2017c; Demoly and André 2022):

– Mechanical strength (E: Young's modulus): "classic" polymeric materials are therefore not suitable for deformation to produce an efficient actuator; indeed, in the literature, we only show deformation, without really associating a mechanical or energetic potential with it (with classic elastomers, a deformation Δy of 0.1 cm for a beam of size 1 cm (thickness and width 0.1 cm) corresponds to an approximate weight of 5.10^{-3} N. For a metal at 100–500 N).

– Quantum efficiency: with single-photon light absorptions, the absorbed intensity follows a decreasing exponential law in simple cases, with greater effects in the vicinity of photon arrival. This is a complex, nonlinear process whose deformations remain delicate. The sharing of incident photons is a factor in the thermal rise of the medium. In practice, we need to consider conditions where thermal aspects can be neglected (small actuators).

– Heat transfer (Fourier's law): polymers are poor conductors of heat. If it is heat that induces physical transformations in the object, the temperature must change in the mass of the material (which explains the choice of light stimulation). From an application point of view, the average characteristic heat transfer time τ is of the order of 15 minutes for a cube of 1 cm on a side.

– Matter transfer (Fick's law): matter transfers are generally between 10 and 100 times slower than heat transfers. Any effects will therefore be very slow. This is what happens, for example, when a deformation is induced by pH, requiring the diffusion of H+ atoms into the bulk material, or the transfer of a solvent within a polymer matrix.

– Manufacturing anisotropy: if an object is produced using stereo-lithography (currently with one photon), the light arriving on the surface to be polymerized is absorbed more at the surface than at depth (André 2017a). The part present towards the surface is therefore more polymerized (or even rigidified) than the areas located deeper down. Under these conditions, the elasticity of materials, whether active or not, depends on spatial parameters.

– Reproducibility, fatigue, breakage: robustness is rarely studied.

– Localization of active materials in the object and spatiotemporal localization of energy stimuli.

– Mathematical control of deformations: the optimal design, response time and local amplitudes of stimuli in the active material, and the overall dimensions of the system for producing energy stimuli need to be studied. Figure 2.14 illustrates one of the difficulties exploiting 4D printing: knowing the properties of the materials under stimulation, how and where to position the active voxels in an inert support material, how to calculate the local flows in time and space of energetic stimuli (advantageously using light?) so that a 3D object moves from one form to another, from one functionality to another?

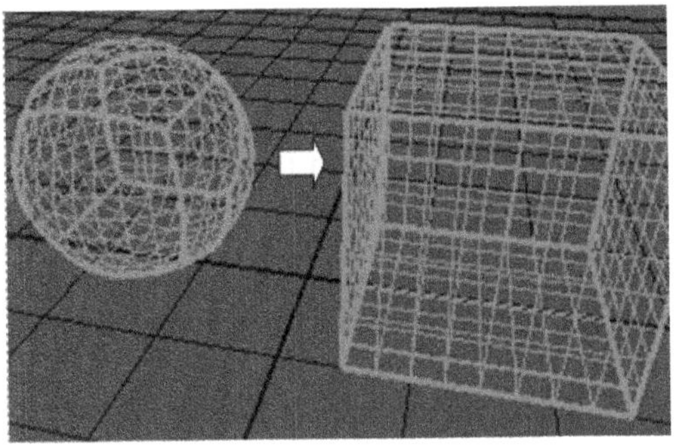

Figure 2.14. *The question of modeling of shape by stimulation, the "possible" case of a structure composed of activatable segments*

These elements define real limits to the development of 4D processes. As long as these locks are not dealt with in a satisfactory manner, only a few specific niches will maintain the activity: for example, for low response times, it would be necessary to produce objects of very small sizes or perhaps to use optical stimulations (process stereo-lithography) or electromagnetic (wire fusion process), for which we know to have time bases adapted to the request. In any case, these technological constraints are lines of research to be carried out if 4D printing is to emerge as an industrial technology.

2.5.1.2. *Breaking the deadlock*

> Let's not mince words. This 'engineering' is a modest tinkering that is much more akin to the empirical procedures of the amateur gardener than to the reasoned knowledge of the agronomist (Rouquette 2016).

> It is a fact that the factory of thoughts is like a weaver's wheel, where a movement of the foot agitates thousands of threads, where the shuttle goes up and down ceaselessly, where the threads slide invisibly, where a thousand knots are formed at once; the philosopher then comes in and shows you that it must be so: the first is that, the second that, therefore the third and fourth that; and that, if the first and second didn't exist, the third and fourth wouldn't exist either. Students in all countries take this reasoning to heart, and yet not one of them has become a weaver (Goethe 2004).

> To each according to his work! Let us generally recognize that freedom is a delicate instrument that needs to be learned, and let's train our children in it; all moral education should be directed to this end (Durkheim 1970).

Koestler (1964) asserts that creative originality does not consist of creating ideas *ex-nihilo*, but rather in combining well-established patterns and structures, by hybridization as it were: the creative act is not a creation in the Old Testament sense. It does not create out of nothing. It discovers, mixes and synthesizes facts, ideas, skills and techniques that already existed. The more familiar the parts, the more astonishing the whole. This is what is observed for the different types of 3D printing. So, after a POC, what is expected is an appropriation of the processes and artifacts to invest them with different meanings and find specific uses. Controversies may arise, but they are usually followed by a process of closure and stabilization; visions will be imposed and reflected in the shape of the objects and their uses (Gaglio 2011).

But where does this leave the scientific (and technological) results of 4D printing? The operation of industries essentially involves forms of foresight that are necessary for identifying markets and tasks, and for controlling and measuring them.

With the possibility of proliferation, creative organizations move away from the principles of predictability and, as a result, recognize the idea of the black box (Birkinshaw et al. 2011). In the case of 4D printing, with its (too) many degrees of freedom, creation integrates a number of inputs, produces outputs, without it being possible to establish a priori a causal link between inputs and outputs: "To put it another way, the recognition of the abundance necessary to all creation means that organizations renounce all forms of traceability, and therefore to any linear vision or quest for optimization" (Paris 2010). The act of creation is a proposition born of an intimate conviction, which needs to be shared through applications. For Paris (2010), the creative economy corresponds to an acceptance of failure (a form of irresponsibility).

However, given the previous comments, the author is not aware of this type of transfer associated with 4D manufacturing! And this important question cannot be the subject of a commentary or an opinion, which would have made it possible to examine whether, the economy of creation being an economy of proposals, there would be an intrinsic value to a proposal. For Paris (2010): "Creation thus appears as one of the last refuges of subjectivity, in a world that is becoming more and more obsessed with rationality".

In this context of uncertainties associated with many degrees of freedom, the wider the zone of investigation and production, the greater the risk of error (or rather wandering) due to the great indeterminacy of each parameter, and therefore the uncertainty in assessing its relevance. Continuing with analogy, we can then define a regulatory mechanism in the resolution system under consideration (whether an individual or a restricted group), whose function is to bring the heuristic process back to an acceptable or supposedly near-optimal trajectory zone, when too great a deviation from it has been perceived (whether rightly or wrongly, of course). With the development of digital technology, companies need to think about the new tools, methods and skills they need to develop in order to organize themselves differently and adapt to their environment. This rethinking could be advantageously associated with transformations in terms of production, based on the selection of value-creating ideas (Kin et al. 2021).

According to Ininou and Loilier (2021),

> The term 'idea' is often defined alternately as a mental representation aiming to solve a new problem already known or to solve a new problem, an intention of action, a proposal carrying incremental adaptations or disruptions major in the design of new products or services, an ability to carry out production that is both new and adapted to the context in which it occurs, a strategic resource, or even as a trigger for the innovation process.

Another criterion with a large number of degrees of freedom!

The selection of ideas thus constitutes a recognition of the general interest of the offer, but then raises a real problem of evaluation, choice and resources for a decision commitment whose risk needs to be estimated. To evaluate, you need to be competent on an open subject, which can raise several questions: if an evaluator has the internal capacity to support a radical proposal, then why call on an outsider since they have mastered the subject? If they are "just" a generalist, how can we link project and evaluation? Awareness of the issues at stake should be present at institutional, corporate and/or societal levels, but given the complexity of the radical proposals that can be made, and the difficulty of perceiving them (before they can be appropriated), the operationalization of changes is bound to be slowed down, despite societies that are open to innovation, which is a paradox. So, let us keep it simple!

In groups, the most immediate indicator of guiding phenomena is naturally linked to the mutual criticism exchanged during the research process. In this respect, we may recall Shaw's (1932) observations: the success of groups in reaching correct solutions can be attributed to the frequency with which an individual's unsatisfactory suggestions are rejected by the others: "Snider, Dekoninck and Culley in real time".

However, insofar as the development of profitable ideas associated with 4D printing involves an interdisciplinary approach, a question must be asked:

> Is it possible to create or develop a group in which all members are genuinely (at least partially) creative, even if originality and talent are not naturally distributed equally, without a leader ensuring its progress and cohesion? If the answer is 'we need a leader', then another question arises: how do we provide this leadership? Should the leader be democratic, autocratic, charismatic, etc.? (Enriquez 2017).

Rather than consensual competence, it is the confrontation of points of view and references that seem best able to bring about the emergence of the desired novelty, provided that divergences are genuinely taken into account (Hall and Watson 1970). After trying to test the spontaneous emergence of creations by freeing the imagination so that spontaneity emerges, the net result is that this activity is purely utopian (at least in the experiments carried out, which were modest enough in number to make generalization possible). "Creativity that would spring forth of its own accord is, in reality, a delusion and a pretense" (Chouvier 2017).

2.5.2. *Gaps between paradigms*

> Only a few artists, writers and scholars escape the cold shower of normalization, the return to order, the loss of the magic of the time when everything was still possible [...] it is in this that all art, all innovative thought is transgression (McDougall 1978).

> Our results suggest that there is an ideal zone of creativity in the period of falling asleep, and that to reach it, individuals must find a balance between falling asleep easily and falling asleep too deeply (Lacaux et al. 2021).

> Interdisciplinarity requires translation and mediation services related to intellectual issues that go far beyond science. In my view, they are philosophical. They have to do with the meaning of knowledge, all knowledge, considered as the experience of the human species, aligned with its will to know and its biological survival (Hamel 2002).

In the 4D creative process, each person must be able to reconnect with "his or her original thaumaturgic power to overcome the prohibitive obstacles that paralyze him or her" (Chouvier 2017). In the author's experience, many people refuse to get involved, claiming incompetence. In a group where their competence can make an important contribution, there can be tension between those who dare and those who refuse to overcome an inhibition towards creation with a feeling of self-deprecation. And yet, some scientists are optimistic when they announce that the limits mentioned for 4D materials will be easily overcome in the future:

> A 'technological paradigm' contextually defines the needs to be satisfied, the scientific principles used for this task, the material technology to be used. In other words, a technological paradigm can be defined as a 'model' of solutions to selected techno-economic problems, based on highly selected principles from the natural sciences, […] together with specific rules aimed at acquiring new knowledge and protecting it, as far as possible, against rapid dissemination to competitors (Haleem et al. 2021)[3].

This is what Dosi (1982) proposed to be compared in particular with scientific paradigms in Kuhn's sense. But, at this stage, 4D printing is still seeking its technological paradigm!

> The engineer's job is to mobilize the technical knowledge at his disposal to devise innovations that represent a compromise acceptable to all stakeholders. Technical perfection is the poison of the economy! You have to be prepared to tinker to satisfy as many people as possible. The resources of technology are such that you can't believe an engineer who assures you without batting an eyelid that there is only one efficient and profitable technical solution, and that you have to choose it (Callon and Lacoste 2011).

Reading these sentences, the lack of mastery of inverse problems in dynamic morphogenesis aspects could (should?) have closed the debate. There is no realistic solution today! What we have shown is that, in order to move forward, we need to find a compromise between systems that are between perfectly causal systems, with the possibility of moving voxels along intentional paths, and a more global vision of

3 Author's translation from the French version of the text.

the morphogenetic type dealing with complex systems. It is largely these difficult situations that we should be moving toward.

NOTE.– Stakeholder theory and strategic management.

Effectuation is based on a robust pragmatic logic because it takes into account the choices made by interested economic agents, having the potential to influence the performance of the company. The effectuation approach is structured around interactions with the environment. This method is iterative, evolutionary and cumulative. "For the effectuation, more than the quality of the product, it is the ability of the entrepreneur to obtain the commitment of a growing number of stakeholders which is the key to the success of the project" (Silberzahn 2013). When it commits, the company has little information on the future of an emerging technology; it can hardly plan its association with the designers.

2.5.3. *Conclusion*

> The hacker attitude promotes free creativity against hierarchical organization, passion against duty, and curiosity against the limits imposed by regulations (Himanen 2001).

> I don't like work. No one likes it, but I do like, in work, that which is an opportunity to discover oneself, I mean our own reality, what we are in our own eyes and not in facade, what others cannot know, because they only see the spectacle and can never be quite sure of what it means (Conrad 1993).

> The first thing is to abandon this sleepwalking mentality, to appear to take seriously the questions, concerns, objections they are accustomed to dismissing as unscientific. The somnambulist has little to do with a 'scientific mind'. It is the product of an institution that defines itself as 'mobilized' by the advance of knowledge (Stengers 2014).

Lepage (2015), largely cynical, writes: "As the domain of the state expands, it becomes more profitable for entrepreneurs to buy political staff rather than buying their customers by offering them what they demand [...] by intervening in economic life in ever more extensive ways, the state itself creates the conditions necessary for increasing manipulation of market laws to the benefit of special interest groups".

All it would take to get ahead would be some good lobbying! It is cynical, and perhaps possible, but the author has never practiced this kind of malpractice...and does not even count on readers to inform him! But for all that:

> As far as the public sector is concerned, the innovation system is no exception to the administrative 'mille-feuilles': 6,000 offices distribute 400 types of aid, mainly supporting employment (a lot) and innovation (little). The reason for this 'sprinkling' is that 'public investors are afraid of failure'. They are not always comfortable with technology and fear unpleasant remarks from their constituents (Battini 2015).

The efficiency of all this administration is therefore very poor!

According to Nelson and Winter (1982), public intervention is focused on financing and carrying out major R&D programs. The state must then implement measures that enable cooperation and synergy between all agents in order to generate resources for innovation and thus improve the country's economic and technological attractiveness (Casadella et al. 2017). However, this funding, which is not covered by specific budget headings, is not considerable, and the return on these investments is likely to be modest!

Furthermore, Alter (2010) has introduced the concept of "dogmatic invention", corresponding to inappropriate inventions that ultimately lead to regression, time wasting and getting bogged down. The image of the idea is tarnished, with the risk that it will no longer be able to recharge its batteries and once again become an innovation, bringing progress to society. At first glance, this is what we can learn from 4D printing today. The notion of innovation systems can be exploited by focusing the analysis cursor on what can induce technological change (technology push model) by orienting research policies (upstream-downstream coupling). The ambition is to mobilize scientific, technical, institutional and entrepreneurial forces, playing on realistic development as an end and a means to the progress enabled by 4D printing, using specific application examples. A number of applications were presented, which do not have to take into account the general shortcomings of 4D technologies, because the applications selected can easily overcome them. Clearly, this is a way of getting society interested in this emerging field.

In this sector, we find ourselves with unpredictable markets, with blurred boundaries, associated with industrial structures that are at least partially non-existent. The presence of designers working on profitable 4D proposals could thus contribute to successful activities, provided they are recognized as such. These are particularly attractive markets, fast-moving and "chaotic", that the pairing of designers and determined industrialists would undoubtedly have the greatest impact in promoting the field (Le Bas 2016). After a (well-chosen) demonstrator meeting an industrial specification, there would be an opportunity to demonstrate the economic interest of 4D technologies and, as a result, drive their promotion. What is more, for the experts at the firms involved in this type of initiative, it is a way of imagining the future products that will guide the organization's medium-to-long-term trajectory.

Design-thinking can provide support for developing innovative capabilities, combining human values, skills and the confidence needed to identify emerging problems and needs together. It contributes to a process of meaning creation, emphasizing imagination and improvisation (Auernhammer and Roth 2021; Rylander Eklund et al. 2022). In this trial-and-error approach, it is not possible to know whether constructs disrupt or support innovation and/or are generalizable.

Tibbits' (2013, 2014) publications are a form of initial promise in terms of applications, not followed by many useful results. Applications of 4D in soft robotics or actuation today do not lead to easily exploitable effects due to the limitations of utilizing the use of adaptive materials (Tang et al. 2017). It is certainly possible to better support interdisciplinarity, to fund risk-taking, to seek out divergent minds, etc. But at the end of the day, the essential question is whether the current situation is inescapable, or whether there are unexplored or little-explored avenues that could satisfy both a robust scientific approach and real applicative openings? And yet, in the fields of flexible robotics and actuation, work close to 4D manufacturing has been in progress for several years.

2.6. Bio-printing

> A writer needs a pen, an artist needs a brush, but a filmmaker needs an army (Orson Welles, quoted by Paris (2017)).

> Man, like every living creature, is constantly thinking but does not know it; the thinking which becomes *conscious* is only the smallest part of it: let's say the shallowest, worst part – for only that conscious thinking *takes place in words, that is in communication symbols*; and this fact discloses the origin of consciousness (Nietzsche 1982).

> The horizon of science remains a formidable tongue in cheek. All scientific institutions are constantly talking about interdisciplinarity, but in reality, the balance of power is merciless. Interdisciplinarity, of course, as long as nothing changes. And like many others, academics and researchers, I've encountered the double difficulty of moving forward without the comfort of disciplinary traditions (Wolton 2013).

> The 'principle of scale relativity' postulates that the fundamental laws of nature must be valid, whatever the 'state of scale' of the reference system. Only scale relationships have any meaning, never an absolute scale. Such a concept also establishes itself as a fundamental principle for 'constraining the description of a non-differentiable space-time' (Chaline 2012).

While bio-printing is the 4D printing of living beings, the first question arises as to the size of the voxels: the smallest size is that of a cell, that is, a size of less than 10 μm. Making a 1 cm^3 bio-construct corresponds to 10^9 voxels! This technically unrealistic situation (the cells must survive) requires the introduction of amplification processes. Otherwise, nothing prevents the voxels from containing a cluster of cells. However, as shown in Figure 2.15, living environments are characterized by the existence of various scales.

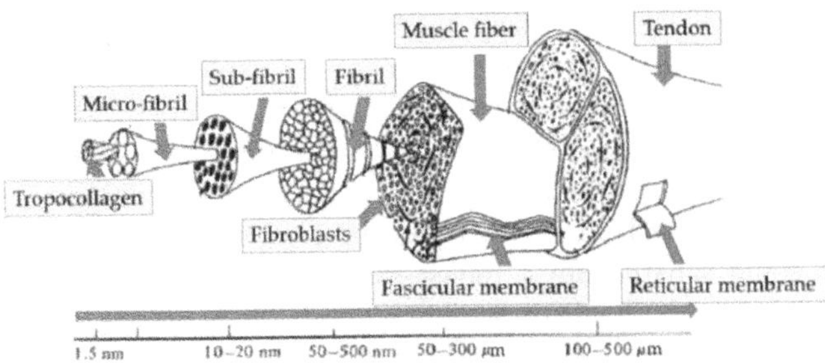

Figure 2.15. *Scale elements to consider (tendon case) (source: Sommer (2016))*

The figure also shows the existence of different materials. There may be others, such as bone and cartilage. Ultimately, there is not just one type of living cell! In short, bio-printing is well understood in principle, but with far greater degrees of complication than 4D printing (which, as we have seen, already has its own difficulties).

2.6.1. *Principle of bio-printing*

> They pretend to train us for society, and we are instructed as if each of us were to spend his life thinking alone in his cell, or dealing with indifferent people on idle subjects (Rousseau 2009).
>
> While disciplinary projects distinguish, privilege and consecrate, interdisciplinary programs combine, solidarize and demystify. It may represent an advanced stage in the secularization of knowledge. The movement of knowledge implies a permanent displacement of borders, or rather, the creation of a cross-border territory (Portella 1992).
>
> Let's not forget that it's only far from equilibrium that mutations are possible, and today transformations, not only possible, but urgent (Mayor 1992).

Bio-printing, according to Peng et al. (2016), has many advantages in principle relative to other biomanufacturing methods such as molding, magnetic assemblies and microfluidic approaches for the following reasons:

– It enables the production of bioprinted tissues that comply with expectations in terms of geometry and (partially) functionality; it exploits the performance of 3D printing.

– It enables the manufacture of porous structures (to which cells can attach themselves) with controlled architecture.

– It allows co-cultures with multiple cell types.

– It enables model systems and various additions.

– It enables the delivery of growth factors and genes.

– It is flexible and can be used to manufacture "made-to-measure" fabrics.

– Integration of the vasculature is possible (Richards et al. 2017).

– Four types of bio-inks are used: hydrogels containing cells, micro-transporters, cell aggregates and de-cellularized matrices.

– Note: nothing is said by these authors about the cell history induced (differential ageing) by the manufacturing process itself: for a long manufacturing time (several hours), the cells printed at the end of the operation will have a different environment to that of the cells deposited at the very beginning, which may have an impact on future cell behavior.

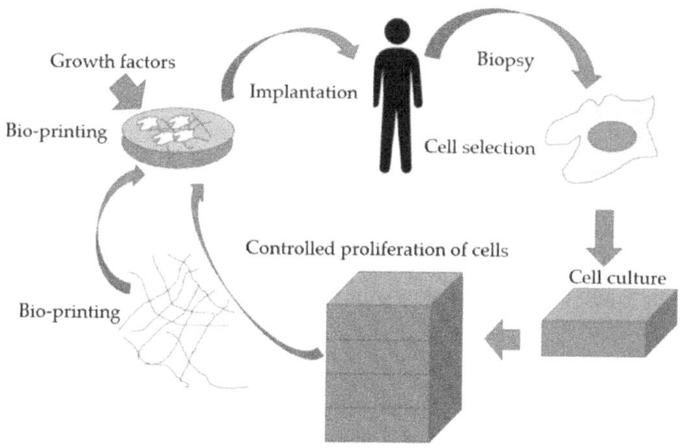

Figure 2.16. *General principle of use ("virtuous circle") of bio-printing*

If it involves the replacement of an organ or tissue, a reasonable number of cells (particularly healthy ones) must be available. Figure 2.16 shows the general principle for human repair using the body's own cells. It requires an amplification process prior to bioprinting, followed by a second process to obtain a suitable cell mass.

Chneiweiss (2012) describes the history of organ transplants as a series of failures until immune compatibility was mastered, with the attendant problems of post-operative treatment. To avoid this major difficulty, a new approach involves the development of organ manufacturing using the patient's own stem cells and bio-compatible materials (a stem cell is defined as a cell with the ability to renew itself identically, without limit, and to be the cell of origin of numerous differentiated cells, capable of forming the structure and supporting the functions of the body's various tissues). The contribution of these differentiated cells in vivo or in vitro (oriented towards the reformation of a specific organ) repairs a damaged organ. Several conditions must be met:

– The cells must be sufficiently differentiated to perform the specific function required.

– They must be able to survive to occupy the space provided by the 3D printing, and stay alive to fulfill the organ's function.

– They must remain in the desired state (is the mechanism used to orient them stable? Are there possible bifurcations or risks of escape?).

– They must be able to establish the necessary functional contacts with their environment (re-establishing cellular network function).

Despite reservations and constraints to be mastered, this is a particularly "exciting" emerging situation, a real scientific and applicative challenge, which concerns the realization of objects livings in bio-materials. Behind bio-printing is the (immodest) idea of being able to model the living world, in the belief that it will be possible to overcome the immense incompleteness of our knowledge of this world. Printing living elements of all kinds is a classic additive manufacturing technique applied to living matter. But it is not so much the technology that is problematic; it is its use with living cells that explores the complexity of nature's phenomena.

2.6.1.1. *Technical point of view*

> We don't have the right to demand that a promise be kept, since to promise is already to bet on time, on the permanence of the subject who will not have fallen into madness or succumbed to death. But to promise is to believe that language commits you and that a truth is at stake in everything that is said, rather in the way and the moment it is

> said, in the intention of the body and the voice, in the contained presence and the surrounding silence (Dufourmentelle 2011).

> The claim of science to dissolve into the anonymity of the mechanical, physical and chemical environment, these centers of organization, adaptation and invention that are living beings. Hence the inadequacy of any biology that [...] seeks to eliminate all consideration of meaning (Canguilhem 1965).

> The fact remains that the construction of a tissue or the differentiation of an organmacroscopic phenomena, must be considered as the integrated result of multiple microscopic interactions due to proteins, and based on their stereospecific recognition properties (Monod 2014).

To achieve ordered cell printing, the cells are prepared using a process based on the same principle as shown in Figure 2.16. This is already a complex biological preparation/purification step, which should leave the cells to be able to develop during the rest of the bioprinting process. The ultimate level involves the use of undifferentiated stem cells, with the conditions in which they are placed determining their cellular future. The idea is therefore simple in principle: to implant cells on a 3D support which will dissolve over time, while the stem cells develop for the desired purpose, that of rebuilding or, perhaps even, increasing human performance. However, there are a number of limitations, such as technological (discussed below), scientific in terms of how to orientate transformation toward such an end, tissue heterogeneity (angiogenesis, vessels, nerves, etc.), changes of scale (from cellular to human compartments), various compatibilities, regulation, ethics, medical training, etc.

From a technological point of view, the “simple” placement of cells does not allow them to survive due to a lack of sufficient nutrient and oxygen flow. A second issue then arises, that is, the need for scaffolds on which the cells will be deposited (for an acceptable answer to this question, we need to choose the right materials and design the scaffolds so that the deposited cells can grow). Depending on the situation, the response levels may vary. In principle, the bio-construct is based on a scaffold – a solid structure – and the deposition of living cells, possibly encapsulated in hydrogels. After maturation, the organ must be able to be implanted and its scaffold structure gradually decomposed, so that the organ can take its place and become vascularized (angio-genesis). To improve vascularization, the design of the tissue includes micro-channels in the structure (allowing nutrients and oxygen to enter and waste to be eliminated).

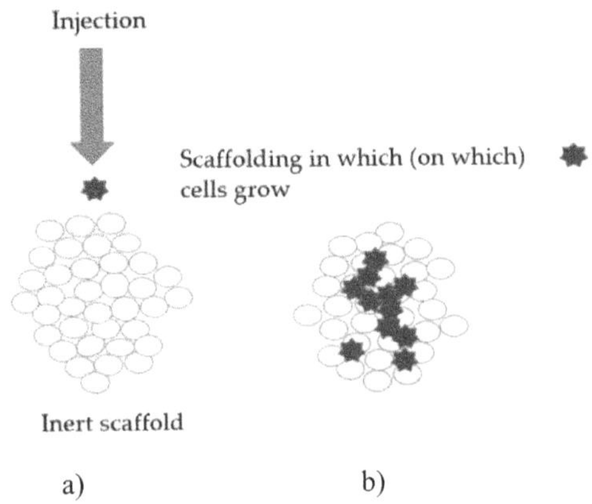

Figure 2.17. *Cell carrier-deposit association in bioprinting*

The generic idea behind bio-printing is shown in Figure 2.17: to place the "right" cells on the "right" scaffolding, so that a living tissue can grow (Melchels et al. 2012; André 2017c). In the name of bio-printing, researchers are striving to build "living bricks" as intelligently as possible, complex voxels that are both highly simplified and controllable, which can then be "stacked" to produce viable tissues that meet functional specifications (and why not more competitive living beings in the future!). In Figure 2.18 from the same source, cell development is operationalized around a flow of matter and energy.

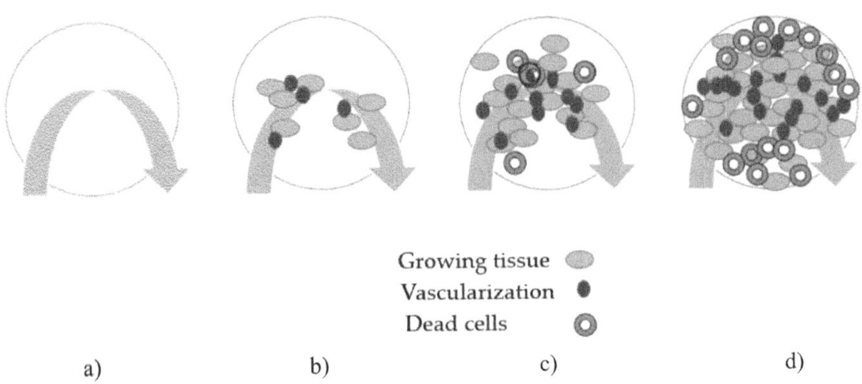

Figure 2.18. *Principle of bio-printing with changes of living printed materials (André (2017c))*

Compared to conventional additive manufacturing techniques, the printing of biological elements adds very high levels of complexity to the processes because it is necessary to know where to position before "intelligently" structuring living or non-mimicking materials that mimic the extracellular matrix, and to control the spatial distributions of different types of cells or biomolecules that can play a role in cell differentiation, growth or apoptosis, etc. What has just been recalled, if we wish to make a comparison with "classic" additive manufacturing processes, is that we add solid additional locks, some of which are conceptual:

– a complex choice of supports for the development of systems using living cells systems;

– cells of variable origin, depending on the purpose;

– evolution of cellular systems, complex functions of spatial, chemical and mechanical environmental factors, interactions with the substrate, various transfers, etc.;

– intermediate cell growth before deposition;

– interdependencies between these three main areas.

Today, it seems inconceivable to master the bio-printing system from knowledge of its constituent parts, hence the understandable tendency toward a global approach as practised by engineers, taking into account the phenomena deemed to be preponderant in the model; but we must therefore move away from the "materialism of promise" practised by some scientists. According to holistic theories, each cell influences the whole under construction, which in turn has a recursive effect on the elementary entity. "There is a sequence of actions and reactions, in other words, we have a system of equilibrium between different forces, rather than a phenomenon that can be reduced to a cause and the effects of that cause" (Pareto 1965).

But then, how can we be sure that we are dealing with the main variables if complex local interdependencies (cells/environment and vice versa) exist? And yet, the objects of bio-printing are composite, systemic, unstable and subject to constraints (some of which are still unknown to us). So, for the author, whatever the scales considered, bio-printing is probably inaccessible to mono-disciplinary approaches alone.

2.6.1.2. *Conceptual viewpoint*

> Engineers hate complexity. I hate emergent properties. I don't want the plane I take tomorrow to have any emergent properties while it is flying (Drew 2010).

> By focusing on the 'useful', we risk compromising the very fruitfulness of the social contribution of science (Guillin 2015).

> This means that a correctly posed problem does not necessarily have a solution (Turing according to Levin (2010)).

The varied-scale approach consists of substituting for a given system, in particular a living system, a different environment presenting "just enough" analogies with the first, in order to be able to infer credible and repeatable conclusions; this is smaller and therefore, in principle, easier to study. What is more, we need to remember that we are always working with a body of knowledge that is largely incomplete, and whose degree of generalizability to larger or smaller scales may be questionable. The idea, however, is to study the control parameters (temperature, various flows, different heterogeneities, spatial distributions, etc.) in an attempt to examine the existence of attractors and their deformation(s) induced by the environment.

> Most of the processes responsible for the functioning of living organisms, at all scales – from the cell to the ecosystem – result from the interactions of tens, hundreds, even thousands of interacting factors: the presence or absence of an ion, a molecule, a nutrient, a gene, the conformation of a protein, etc. Deepening our understanding of them therefore requires mastery of tools capable of representing and analyzing these complex interactions. This is what is known today as 'network biology' (Vandeginste 2012).

Biological growth laws are associated with numerous nonlinear dynamic phenomena due to the wealth of scenarios already observed. Taking a cross-disciplinary approach to this field (involving all the disciplines concerned with the notion of convergence) is clearly a key element in any attempt to move the subject forward. Remember that in linear systems, parameter modifications lead to quantitative changes, but do not modify the behavior (steady state) of the system. In nonlinear dynamical systems (NLDS), a small variation in certain parameters, known as control parameters, under well-defined conditions, in the vicinity of a critical value, can bring about a complete change in the system's equilibrium behavior. This is known as a bifurcation. Figure 2.19 shows such a "fork bifurcation".

> Thus, what appears 'simple' a priori is in fact the result of a concerted evolution of various characters, leading to a real optimization of means. It is therefore, from a biological point of view, something very complex, and the initial vision of the ancestral simple disappears through this reversal due to evolutionary reasoning (Le Guyader 2008).

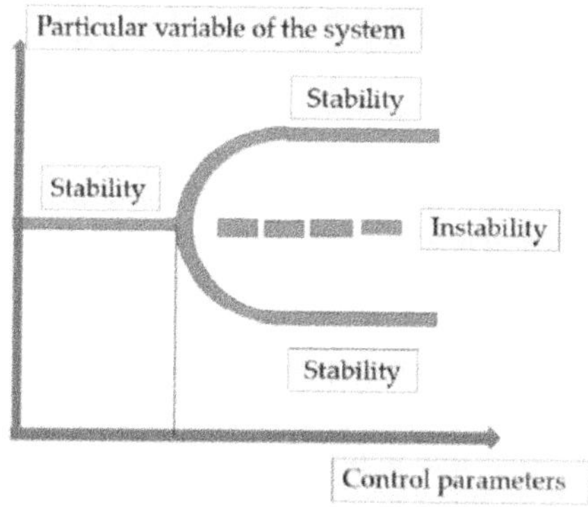

Figure 2.19. *Fork bifurcation*

Figure 2.20, taken from Delignières (2015), recalls, in a two-dimensional representation, situations associated with possible bifurcations (reaching different states from the same behavior law with multi-stable states).

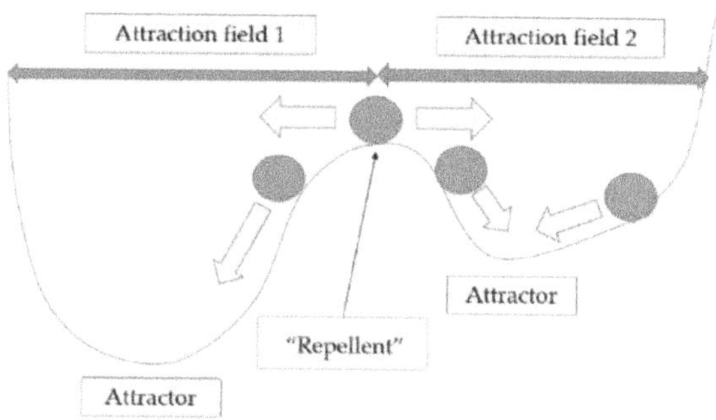

Figure 2.20. *Schematic representation of a landscape of attractors*

In this figure, the stability levels are located in basins of attraction, while the repellents correspond to zones of maximum instability; in the figure, the ball can move either to the right or to the left, depending on the very low-amplitude impulse it receives. The essential characteristic of the attractor is the stability of the order parameter, with a strong tendency toward reproducibility. According to Delignières (2015), bifurcations are heralded by critical fluctuations, that is, a significant evolution of the order parameter in the vicinity of the transition (as in a change in the orientation of molecules in an organized system as soon as a transition temperature is exceeded). Ultimately, it may be possible to find an approach that can be applied if the number and quality of basins of attraction are not significantly affected by these changes.

To put this another way, is it possible to show the possibility of having, as Thomas et al. (2004) propose, "chaotic labyrinths" that constrain complexity? For bio-printing, starting from a given position of the cells in a given environment, does the evolution of the system always converge toward the same solution? In any case, the information generated by the experimental approach to be undertaken should provide data (if possible, robust data) on systems' livings and their evolutionary dynamics, including the possible existence of basins of attraction. If only one exists, it is possible to think that the evolving system can be approached using a deterministic model. But will this be a general case? If we refer to a recent article by biologists (Olle-Villa et al. 2016) that does not illustrate this potentiality of having a single bio-manufacturing pathway in a biological system, there is room for doubt (see Figure 2.21).

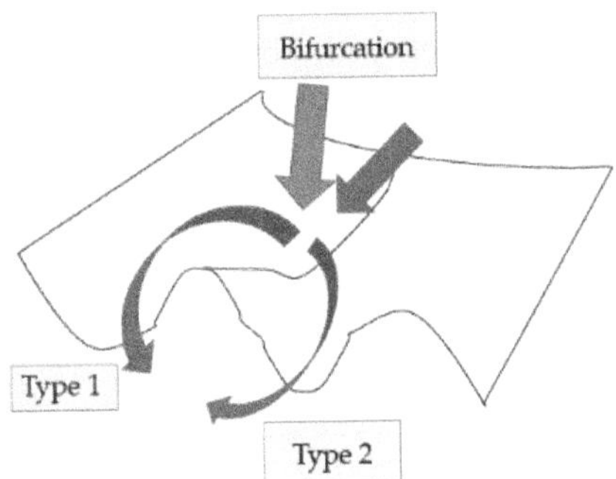

Figure 2.21. *Bifurcations in a biological system*

The challenge would therefore be to carry out large-scale scientific research in order to build up a knowledge base that would enable us to assess the causal credibility of the bio-printing project with regards to organ printing (and which organs). To take an example, if a system made up of one voxel (or several) can undergo a bifurcation that cannot be controlled by external parameters, how can we credibly envisage the relevance of an operation that might one day lead to a transplant?

2.7. Discussion

> It is the most radical way of annihilating all discourse to isolate each reality from any attachment to the rest; for it is through the mutual combination of forms that discourse is born to us (Piaget 1967).

> In modern physics and biology, problems of organized complexity arise everywhere, i.e. involving the interactions of a large number of variables; they call for new conceptual tools (von Bertalanffy 1973).

> In general, we simplify by means of a forced choice: continuous or dis-continuous, analysis or synthesis, excluding the third party. God or the devil, yes or no, with me or against me, of two things only one. But complexity beckons to the side of reality, whereas dualism calls for battle, where new thought dies, and the object disappears. Dualism serves to properly define niches where fighters who lack courage settle, balanced for the long haul (Serres 2013).

In order to stimulate critical reflection on the guiding ideas for taking this subject further, it should be useful to keep in mind the founding intentions, guiding concepts and methodological, paradigmatic, anthropological, epistemological and ethical presuppositions of each discipline. At the same time, however, we must humbly confine ourselves to a task of clarification. Technologies are only visible elements of a larger whole, whose physical and conceptual constituents only come together as a result of a shared history (Tinland 1994). In view of the above, if we wish to compare bio-printing with other with "conventional" additive manufacturing processes, we are adding additional barriers, some of which are conceptual (Godin 2015):

– a complex choice of resorbable supports (scaffolding) for the development of systems using living cells;

– for example, the biological phenomena involved in the development process are very numerous and their laws are much less well known than the elementary laws of physics;

– cells of variable origin, depending on the purpose;

– interactions with the environment are complex and continuously modulate cell growth;

– changes in cellular systems, complex functions of spatial, chemical and mechanical environmental factors, interactions with the support, various transfers, etc.; the levels of organization are numerous and closely intertwined, making it extremely difficult, even arbitrary, to break down such a system into simpler and more independent elementary subsystems;

– does the change of scale respect the principles of self-similarity?

– is the intermediate growth of cells before deposition detrimental to the objective?

– initial conditions are not defined;

– interdependencies between these main areas.

At first, it is easy to follow Besnier and Perriault's (2013) critical comments when, in many organizations that fund research, it is possible to hear a discourse supporting disciplinary hybridization as the key to scientific and technical innovation. But, at the same time, we can see that interdisciplinarity is only very moderately supported for a variety of reasons: the "cross-disciplinary" skills of expert evaluators, the consequent risk-taking involved, the choice of scientists from different disciplines, the quality of the leader, the need for a high degree of freedom etc.

With bio-printing, you have to involve several disciplines (it is hard enough to exchange ideas and act in pairs!) without being sure of the result. "Tell yourself a priori that something is going to change. You set yourself an objective, you launch proposals, you sow various seeds, and you know that time will bring an answer" (Moreno 1990). In this form of research, it may involve the use of diverse perspectives from several disciplines, but, obviously, it is best conducted as an interdisciplinary approach (Czajkowski et al. 2016). Irrespective of this strong constraint of integrating of disjointed knowledge, working on simple biological systems such as vascularization, it is possible to produce bio-constructs with cartilage, seeming to show that once this step has been taken, it will be possible to attack more complex biological systems. So, certain proposals give the impression of being consensual, even if they are validated by peers; however, this is precisely because their role is to build consensus between peers and to prepare newcomers on this type of subject (idea of disciplinary socialization according to socialization according to Raynaud (2021)).

In the current French context, where many good proposals are not accepted due to a lack of available funding, it is easy to understand the position of the experts who evaluate the dossiers, choosing on the basis of scientific quality and the best possible investment of funding. With 3D printing (stereo-lithography), invention was easy (just one person, a few days for demonstration) and it is not realistic (apart from a significant result, the result of a possible serendipity) to be able to extend what was produced in 1984 to interdisciplinary convergence actions in the field of bio-printing. Dacheux (2013) reminds us that the so-called "hard" sciences seek to establish laws, regularities, to eliminate the irregular and the unique.

In biology, each living being is unique. Taleb (2011) shows the importance of "the power of the unpredictable" (rare events such as bifurcations have, for example, a prominent role in the fate of a bioprint). In order to achieve this objective of developing an epistemology adapted to this complex system, can we agree to popularize our knowledge (so that it is understood by everyone) at an interdisciplinary meeting on such a theme with an applicative aim? It would be worthwhile for each discipline to present a state of low complexity (Naudon 2013).

From this minimalist pooling, could we go any further (see, for example, Klein 2010)? The debate is open!

Le Crosnier et al. (2013) consider that, in principle, “the sciences produce a large number of hybrid objects, both artifacts or theories that fall within the scope of validation by scientific communities, but also products or methods that are widespread in society”. This interweaving of science and society can give rise to sociotechnical controversies that can drive the subject forward. Two questions may be associated with this interesting option:

– Who are the scientists who would be willing to commit to such a theme, extended to society as a whole (problems of evaluation and careers)?

– Who would pay for such an experiment?

As far as the progress of bio-printing is concerned, given the complexity of the subject, is it not fair to say that this type of topic has been given more modest consideration than others of obvious social utility? But the context of responsibility has two sides: the researcher and the client (Haven et al. 2020). So, rather than to attempt a synthesis in the form of a popularization, Bourdin (2022) suggests engaging in a gradual process “in which the partners share, not the description of observations or the results of hypothesis tests, but links between ideas, which they gradually and reciprocally consolidate”. This author proposes a commitment on three axes, which are:

– a draft understood by other disciplines;

– defining the concepts relevant to the problem to be addressed;

– adjusting notions to each other in the context of the whole, while preserving their “acceptability” by their original discipline. Once the framework has been sketched out, it can be broken down into tasks.

This procedural approach may be interesting, but it cannot easily take into account the bifurcations and other complexities associated with bioprinting. What can be useful in this type of approach is the possibility of narrowing the scope of the domain to be addressed. More precisely, Pineo et al. (2021) propose the following approach (somewhat theoretically, not experimentally validated):

– a co-learning phase which recognizes the distributed nature of knowledge generation and learning between partners;

– a pre-development produced to establish a shared vision and working methods comprehensible to all parties; but these exchanges can only be understood “in relation to the subversive or legitimizing role they play in the host discipline” (Grenier and Lepetit 1989);

– reflection on the improvement processes and results in line with emerging priorities considered relevant, with genuine interaction, that is, reciprocity in exchanges, so that there is a total mutual enrichment;

– conceptualization corresponding to the pooling of knowledge from different disciplines, avoiding a juxtaposition of viewpoints with no overall coherence;

– implementation of the proposal.

> From different horizons comes the idea that order, disorder and organization must be thought of together. The mission of science is no longer to banish disorder from its theories, but to deal with it. It is no longer to dissolve the idea of organization, but to conceive and introduce it to federate fragmented disciplines (Romian 2000).

Despite the considerable potential for applications, for the time being, we are still in the exploratory phase of 3D printing of living matter, far ahead of a new techno-industrial paradigm. New 3D technologies already cover the field of prosthetics, dentistry, etc., because they are "classic" 3D printing with certain adjustments. So, the transformation of value chains, in which new customer–supplier relationships are at stake, is possible. With bio-printing, manufacturers are waiting to combine technological skills with "business" skills of scientific origin, in order to bring the technology closer to its potential applications, particularly in medicine.

2.8. Conclusion

> It takes two to invent. One forms combinations, the other chooses, recognizes what he wants, what is important to him in the ensemble of the first's products. What we call 'genius' is much less the act of the former – the act of combining – than the readiness of the latter to understand the value of what has just happened, and to grasp that product (Valéry 1957).

> We shouldn't worry about whether what we're doing is science, but whether what we're doing is likely to touch on a truth. You don't have to choose between art and science. You choose art in the sense of doing what you have to do well to try to show what you think is a truth, rather than knowing whether you are responding to modes of estimation that will make people recognize that what you are doing is science. We must not start from a system of opposition, nor from the idea that there is art (or literature) on one side and science on the other (Rancière 1994).

> We don't have to know in advance where to start. We don't know what makes up our world, who acts in it, or what hardships they put each other through. Nor do we know what is important and what is insignificant, and what causes the shifts we observe around us (Latour 1984).

> Creativity presupposes preparation, even unconscious preparation, a cultural environment or, if you prefer, a terrain that only in-depth work can provide. It should be noted, moreover, that one of the conditions of creativity is not only deepening, but also its corollary: broadening. Voltaire's formula is true: 'Our soul expands as it fills'; but it finds its truth in a humanist attitude, which is the will to understand others and to open up to them. Deepening without broadening may lead to erudition, but never to culture (Cramarégeas 2021).

If a demonstrator is to be successful in convincing an industrialist, this preparation for creativity must be extended to other fields, in particular through interdisciplinary approaches. However, what this book shows is the existence of limits apart from the simplistic (and successful!) principle of additive manufacturing. "The unexpected [...] exacerbates the feeling of powerlessness, which leads to inaction [...]. Forecasting doesn't cancel out the panic effect, it prepares for it." (Lavis 2021). These delays or questionings have their origins in the nature of the materials (4D) or deep-rooted epistemological difficulties (bio-printing) and yet, the number of degrees of freedom is very high in both cases, market appetite is high, etc. It is therefore important to try and understand why technologies that seem particularly attractive are not being developed. For example, Altos Labs is a start-up that proposes to rejuvenate cells and, as a result, claim a form of immortality. It is a 3 billion dollar dream (Pingeot 2022). Not to mention if, on top of that, we are able to produce bio-tissues and organs!

Today's common identification of scientific research with industrial innovation can be dangerous because only a fraction of scientific advances find technical applications, which are both limited and unpredictable; this approach perhaps favors less innovative research, offering short-term prospects of results, but which have the advantage of being operational.

As a result, material resources (see projects supported by the EU and ANR – French National Research Agency) are focused only on major fields with immediate potential applications, and are stagnating, or even regressing, in speculative research fields that are apparently less promising, but just as promising in terms of innovation. What we see is that new technical innovations are very rarely accompanied by processes of democratization and ethical questioning of their value (Lechevalier 2019). In France, it is possible to think that the industrial world is unburdening itself a little too much of its own responsibilities when it comes to applied research, where the public authorities take on investments from which the private sector will reap the benefits. But, in this sense, the steering of finalized research by this sector is in its hands. It is the vision of a profitable future that governs academic action. From the perspective of productive development, this is a short-sighted policy, as it risks blocking the possibility of useful, radically innovative discoveries in the medium term (even in fields considered non-strategic). Above all, such an approach risks locking scientific knowledge into an instrumental, subordinate role, to the detriment of the intellectual and cultural values our society prides itself on (according to Lévy-Leblond 2015). Indeed, we might legitimately ask whether the private sector is giving back to the public sector the benefits it derives from the relationships it forges with the academic sector.

On the academic side, scientific creation is only possible with the cooperation of a principle of profusion, imagination and disorder, and a principle of rigor, method, criticism and regulated order; these have to work together, which is not without its tensions. Project management therefore requires real freedom of action (Sarooghi et al. 2015). But creative mediation is about creating links between people or groups that did not exist before, links that will benefit them both. From a "tactical" point of view, for an operation carried out in the form of a project to be productive in a rather individualistic culture, it naturally requires a combination of vision, scientific and technical capacity to take action, support from the hierarchy, time management and a flexible organization. The art of "combining" skills is an important factor in the success of an interdisciplinary research operation, as it allows for creativity, the beginnings of which are always very fragile and fleeting.

It is also important to remember that the more uncertain and turbulent a project's environment (that is, risky), the more autonomous the project team must be in terms of process, organization and objectives. This "normal" situation, investing in the complexity of the systems to be studied and/or created, requires teams from diverse backgrounds to engage in concertation of viewpoints and experimentation, going as far as the demonstrator, whenever possible; the only one likely to enable measurable progress through feedback and the emergence of scientific barriers that may or may not be overcome (Anisov 2020).

A classic difficulty lies in the non-flexibility of thinking and acting modes, relying solely on deductive logic without taking into account the compromises that (still) enable progress. The principles of creating autonomous project teams with an end in view need to be developed and tested in a few cases (with feedback).

Cowen (2011), in his book *The Great Stagnation*, notes at the heart of his argument that economic growth in the United States and European industrial countries is suffering from slowed growth caused by the declining rate of innovation. The innovations of the early 20th century had exhausted their effects on economic growth. What is more, although creativity is advocated, it is "on the loose contradictory to the unsubdued essence of any creative process. So much so that we could speak of an 'ideology of creativity' that drives many companies today, for narrowly economic and instrumental purposes" (Amado et al. 2017). A distinction needs to be made between creativity and creation, with creativity coming preferentially from the academic world.

However, the invention of a new marketable form cannot be based solely on a possible proliferation of ideas (criteria of efficiency, productivity and profitability). What is central to creation is the establishment of links that bring about a genuine restructuring around realistic products, with compromises aimed at application (Anzieu 1971). As soon as the context corresponds to the instantaneous effort to be financed in relation to a profitable economic result, companies do their best to keep an eye on science and technology, but are not involved in high-risk research.

The proposals presented above separate the world of science into two useful categories for society: those who understand so that others do, and those who support a creative mess factor of societal initiatives. In this second domain, the creative processes must "intelligently" articulate times of divergence (of bisociation, according to Koestler (1964), in which the action is opened up to multiple proposals), and times of convergence, because it is good in practice to engage in a project, generally interdisciplinary, of societal utility.

How then, in a context where France no longer knows how to innovate very well, can we nurture this fertility and agitation for action that is open to society's needs, open to partners, to giving ideas time to develop and mature, and above all to ensuring that the project comes to fruition (finally emerging from the "valleys of death" syndrome)? The strong interaction between knowledge-building dynamics, creativity and innovation is essential for a successful application, even if it considers different contexts for different types of industries and companies (Parmentier et al. 2021). In today's world, responsiveness seems to have become, with digital technologies, an almost absolute valuation criterion.

> While we must avoid the trap of 'technological determinism', the fact remains that digital technologies provide the conditions for the possibility of completely new modes of instantaneous interaction, which are quite notoriously overturning both our ways of being with ourselves and with others (Chardel and Dartiguepeyrou 2018).

The field of possibilities is particularly vast, whatever the technology under consideration (3D, 4D, bio-printing). So, if it is possible, it is, in the mind of the person who wants it, achievable! In technological fields, responding to a request in an open field can result in the creation of unprecedented situations, where what you want to produce may show abnormal behavior in a system you had not foreseen beforehand. And that is where the problem lies. Indeed, the remarks made above imply the ability, with confidence, to make flexible choices in terms of targets, organization, timing, etc.; and at the same time to get away from the old "romantic" myth of the solitary creator. If it is up to the academic structure to detect talent (what talent?), it is important that the activity does not result in a formatting factor that ensures quality, but conforms to production; or, on the contrary, not to leave a useless dream behind. It is a question of thinking responsibly about how to organize divergent thinking, openness, and even a certain scientific nomadism, and how to develop skills so that the hierarchy can benefit from them in terms of its dynamics and legitimacy of open action on society. Will we then be able to achieve a certain coherence in scientific and/or technological innovation by:

– taking into account the socio-cultural universe it covers?

– respecting the heterogeneity, or even the disjunction of the knowledge gathered, inaccessible to disciplines alone?

– taking into account the heterogeneity of the audiences concerned?

– reinforcing the desire to create and undertake together?

This is certainly a major undertaking, but it has more to do with organizational aspects and responsibilities (at all levels) rather than the need (always legitimate, of course) for an overall increase in funding and positions. The comments made above mean that research organizations need to be places where social ties are forged in ways other than by the simple juxtaposition of opportunities or mechanical cooperation. If there is to be a cultural break, it needs to be carefully thought through. We need to rediscover Wiener's reasoning (1997): "that of the engineer whose gaze does not stop at material facts, but extends to human ones".

What is more, the researcher will develop an even stronger bond with the organization in which they work, if the latter has tried to do the same. In other words, the concern for any institution, which is characterized in every individual by loyalty, attachment, belonging and a job well done, can only be achieved if the institution itself has a concern for people. However, in the concerted activity that must be at work, the question naturally arises of isolated research, a form of scientific autism in a state of breakdown, which is totally separated from reality: "the fact of not having the capacity to step outside oneself because there is nothing to expect from the world" (Castel and Haroche 2001).

2.9. References

AFPR (2015). Guide à la mise en place du prototypage rapide [Online]. Available at: http://code80.net/RP2/content/sourcing/guide/info/contexte.htm.

Aghion, P., Cohen, E., David, B., Gigout-Magiorani, T. (2021). Pour une nouvelle politique industrielle après le Covid-19 [Online]. Available at: https://legrandcontinent.eu/fr/2021/06/04/pour-une-nouvelle-politique-industrielle-apres-le-covid-19/.

Ahmed, K., Shiblee, N.I., Khosla, A., Nagahara, L., Thundat, T., Furukawa, H. (2020). Review – Recent progresses in 4D printing of gel materials. *Journal of Electrochemical Society*, 167, 037563.

Akrich, M., Callon, M., Latour, B. (1988). À quoi tient le succès des innovations ? *Annales des Mines*, 11, 4–29.

Aladad, H. (2009). Conception du système de fabrication de pièces mécaniques en grand série : formalisation de la configuration géométrique (enveloppe) et cinématique de Machine-Outil Reconfigurable (MOR). Thesis, ENSAM – Arts et Métiers ParisTech, Paris.

Alapetite, M. (2014). Ouverture. In *Créativité et inventivité en institution*, Delion, P. (ed.). Erès, Paris.

Alter, N. (2010). *L'innovation ordinaire*. PUF, Paris.

Alvarez-Pereyre, F. (2003). *L'exigence interdisciplinaire*. MSH, Paris.

Amado, G., Bouilloud, J.P., Lhuilier, D., Ulmann, A.L. (2017). Introduction. In *La créativité au travail*, Amado, G., Bouilloud, J.P., Lhuilier, D., Ulmann, A.L. (eds). Erès, Paris.

André, J.C. (2017a). *From Additive Manufacturing to 3D/4D Printing 1: From Concepts to Achievements*. ISTE Ltd, London, and John Wiley & Sons, New York.

André, J.C. (2017b). *From Additive Manufacturing to 3D/4D Printing 2: Current Techniques, Improvements and their Limitations*. ISTE Ltd, London, and John Wiley & Sons, New York.

André, J.C. (2017c). *From Additive Manufacturing to 3D/4D Printing 3: Breakthrough Innovations: Programmable Material, 4D Printing and Bio-printing*. ISTE Ltd, London, and John Wiley & Sons, New York.

André, J.C. (2019). *Industry 4.0: Paradoxes and Conflicts*. ISTE Ltd, London, and John Wiley & Sons, New York.

André, J.C. (2020). Recyclage des déchets d'impression 3D. *Techniques de l'Ingénieur*, BM 7975.

André, J.C. (2021). Bio-printing – De l'organe à la médecine personnalisée, entre résultats et promesses. *Techniques de l'Ingénieur*, RE268 v2.

André, J.C., Le Méhauté, A., De Witte, O. (1984). Dispositif pour réaliser un modèle de pièce industrielle. French Patent No. 84 11 241.

Andreoni, A. and Gregory, M. (2013). Why and how does manufacturing still matter: Old rationales, new realities. *Revue d'économie industrielle*, 144, 21–57.

Anisov, A.M. (2020). Idea of the proof. *RUDN Journal of Philosophy*, 24, 228–243.

Anzieu, D. (1971). L'illusion groupale. *Nouvelle revue de psychanalyse*, 4, 73–94.

Arthur, W.B. (2005). The logic of invention. SFI Working Paper [Online]. Available at: http://sfi-edu.s3.amazonaws.com/sfi-edu/production/uploads/sfi-com/dev/uploads/filer/32/45/324518c2-e7fe-4243-8c5f-76b64d68a36a/05-12-045.pdf.

Assmus, A. and Haeussler, C. (2017). At the interface between invention and innovation: Increasing success rates in clinical trials. *Academy of Management Annual Meeting Proceedings*, 1–1.

Auernhammer, J. and Roth, B. (2021). The origin and evolution of Stanford University's design thinking: From product design to design thinking in innovation management. *Journal of Product Innovation Management*, 38, 623–644.

Battini, P. (2015). La problématique de l'innovation en France. *Vie & sciences de l'entreprise*, 200, 119–141.

Baudrillard, J. (1990). *La transparence du mal*. Galilée, Paris.

Baudrillard, J. (2002). La violence de la mondialisation. *Le Monde diplomatique* [Online]. Available at: https://www.monde-diplomatique.fr/2002/11/BAUDRILLA-RD/9608.

von Bertalanffy, L. (1973). *Théorie générale des systèmes*. Dunod, Paris.

Besnier, J.M. and Perriault, J. (2013). Introduction générale. In *Interdisciplinarité : entre disciplines et indiscipline*. CNRS, Paris.

BFF (2022). L'impression 3D, comment allier production industrielle et réduction des déchets en faisant des économies [Online]. Available at: https://www.forbes.fr/brandvoice/limpression-3d-comment-allier-production-industrielle-et-reduction-des-dechets-en-faisant-des-economies/.

Birkinshaw, J., Bouquet, C., Barsoux, J.L. (2011). The 5 myths of innovation. *MIT Sloan Management Review*, 52, 43–50.

Bonasso, S.G. (2007). Inquiry, discovery, invention, and innovation – The personal experience of technology generation and transfer in engineering and scientific research. *Leadership and Management in Engineering*, 7, 141–150.

Bourdin, S. (2022). Quelle acceptabilité sociale des technologies disruptives ? [Online]. Available at: https://theconversation.com/quelle-acceptabilite-sociale-des-technologies-dis-ruptives-173337.

Burger-Helmchen, T. (2013). *The Economics of Creativity: Ideas, Firms and Markets*. Routeledge, London.

Bush, V. (1945). The endless frontier, report to the president on a program for postwar scientific research [Online]. Available at: https://www.nsf.gov/od/lpa/nsf50/vbush1945.htm.

Buttin, N. and Saffré, B. (2016). *Activer l'économie circulaire*. Eyrolles, Paris.

Caillat, F. (2014). *Foucault contre lui-même*. PUF, Paris.

Callon, M. and Lacoste, A. (2011). Défendre l'innovation responsable. *Debating Innovation*, 1, 5–18.

Canguilhem, G. (1965). *La connaissance de la vie*. Vrin, Paris.

Carlsson, S.A. (2007). Developing knowledge through is design science research: For whom, what type of knowledge, and how. *Scandinavian Journal of Information Systems* 19, 75–86.

Carlsson, S.A (2010). Design science research in information systems: A critical realist approach. In *Design Research in Information Systems: Theory and Practice*, Hevner, A. and Chatterjee, S. (eds). Springer, New York.

Carroll, J.M. (2003). *HCI Models, Theories and Frameworks: Toward a Multidisciplinary Science*. Morgan Kaufmann Publishers, San Francisco.

Carroll, J.M. (2016). Theorizing the unprecedented. In *Theory Development in the Information Sciences*, Sonnewald, D.H. (ed.). University of Texas Press, Austin.

Casadella, V., Tahi, S., Temple, L. (2017). Créativité et politiques d'innovation. *Innovations*, 53, 5–12.

Castel, R. and Haroche, C. (2001). *Propriété privée, propriété sociale, propriété de soi*. Pluriel-Fayard, Paris.

Chaline, J. (2012). La relativité d'échelle dans la morphogenèse du vivant : fractal, déterminisme et hasard. *Annales de chirurgie plastique esthétique*, 5, 517–526.

Chapouthier, G. (2008). Qu'est-ce qu'un biologiste aujourd'hui ? *Pour la science*, 366, 30–33.

Chardel, P.A. and Dartiguepeyrou, C. (2018). Être, temps et différences : pour une approche différencialiste du temps à l'ère numérique. In *@ la recherche du temps – Individus hyperconnectés, société accélérée : tensions et transformations*, Aubert, N. (ed.). Erès, Paris.

Chenavaz, R. (2012). Dynamic pricing, product and process innovation. *European Journal of Operational Research*, 222, 553–557.

Chneiweiss, H. (2012). *L'homme réparé ; espoirs, limites et enjeux de la médecine régénérative*. Plon, Paris.

Chouvier, B. (2017). Groupalité et processus sublimatoire. In *La créativité au travail*, Amado, G., Bouilloud, J.P., Lhuilier, D., Ulmann, A.L. (eds). Erès, Paris.

Christensen, C.M. (1997). *The Innovator's Dilemma: When New Technologies Cause Great Firms to Fail*. Harvard Business School Press, Boston.

Christensen, C.M. and Raynor, M. (2003). *Innovator's Solution: Creating and Sustaining Successful Growth*. Harvard Business School Press, Boston.

Cohen, W., Nelson, R., Walsh, J. (2003). Links and impacts: The influence of public research on industrial R&D. In *Science and Innovation: Rethinking the Rationales for Funding and Governance*, Geuna, A., Salter, A., Steinmuller, E. (eds). Edward Elgar, Cheltenham.

Conner, B.P., Manogharan, G.P., Martof, A.N., Rodomsky, L.M., Rodomsky, C.M., Jordan, D.C., Limperos, J.W. (2014). Making sense of 3-D printing: Creating a map of additive manufacturing products and services. *Additive Manufacturing*, 1(4), 64–76.

Conrad, J. (1993). *Au cœur des ténèbres*. Gallimard, Paris.

Corazza, G. (2015). Potential originality and effectiveness: The dynamic definition of creativity. *Creativity Research Journal*, 28, 258–267.

Cornu, J.M. (2008a). *Nouvelles technologies, nouvelles pensées ? La convergence des NBIC*. FYP, Limoges.

Cornu, J.M. (2008b). Des technologies convergentes… Vers quoi ? [Online]. Available at: http://alltrends.over-blog.net/article-25668522.html.

Cornu, J.M. (2009). La convergence des sciences (1/3) : un choix de société [Online]. Available at: http://www.internetactu.net/2009/05/25/la-convergence-des-sciences-13-un-choix-de-societe/.

Cowen, T. (2011). *The Great Stagnation: How America Ate All the Low-Hanging Fruit of Modern History, Got Sick, and Will (Eventually) Feel Better*. Dutton, New York.

Cramarégeas, F. (2021). L'interdisciplinarité dans l'enseignement secondaire en France (1970-2018) : une légitimation contrariée dans les discours, les configurations institutionnelles et les pratiques enseignantes. Thesis, Université de Bordeaux, Bordeaux [Online]. Available at: https://tel.archives-ouvertes.fr/tel-03483282/document.

Czajkowski, S.M., Lynch, M.R., Hall, K.L., Stipelman, B.A., Haverkos, L., Perl, H., Scott, M.S., Shirley, M.C. (2016). Transdisciplinary translational behavioral (TDTB) research: Opportunities, barriers, and innovations. *Science Translational Medicine*, 6, 32–43.

Dacheux, E. (2013). Redécouvrir les liens entre science et anarchie pour penser l'indiscipline du chercheur et sa nécessaire responsabilité. In *Interdisciplinarité : entre disciplines et indiscipline*. CNRS, Paris.

Damoah, I.S. and Botchie, D. (2020). Do-it-yourself (DIY) laboratories and science, technology, and innovation (STI): Trends, implications and future research. *Technology Analysis & Strategic Management*, 33, 1267–1280.

Delignières, D. (2015). Systèmes dynamiques et morphogénèse [Online]. Available at: http://didier.delignieres.perso.sfr.fr/Supports-doc/systemes_dynamiques_et_morphogenese.pdf.

Demoly, F. and André, J.C. (2022). *4D Printing 1: Between Disruptive Research and Industrial Applications*. ISTE Ltd, London, and John Wiley & Sons, New York.

Despeisse, M., Yang, M., Evans, S., Ford, S., Minshall, T. (2017). Sustainable value road-mapping framework for additive manufacturing. *Procedia CIRP*, 61, 594–599.

DIRECCTE (2014). L'impression 3D : état des lieux et perspectives [Online]. Available at: http://www.centre.direccte.gouv.fr/IMG/pdf/L_impression_3D_etat_des_lieux_et_pers-pectives_-_DIRECCTE_Centre_en_partenariat_avec_CCI_Centre_-_10_decembre_2014-2.pdf.

Dortier, J.F. (2015). Innover… contre les idées reçues. *Les grands dossiers des sciences humaines*, 38(1).

Dosi, G. (1982). Technological paradigms and technological trajectories: A suggested interpretation of the determinant and direction of technological change. *Research Policy*, 11, 147–162.

Doualle, T., Gallais, L., André, J.C. (2021). 3D-printing of silica glass through multiphoton polymerization process. *Optical Letters*, 46, 364–367.

Downes, L. and Nunes, P.F. (2013). Big-bang disruption. *Harvard Business Review*, 91, 44–56.

Drew, E. (2010). Symposium on opportunities and challenges in the emerging field of synthetic biology. *OECD and Royal Soc. Ed.* [Online]. Available at: https://www.oecd.org/sti/biotech/45144066.pdf.

Dubois, M.J.F. and Brault, N. (2021). *Manuel d'épistémologie pour l'ingénieur*. Éditions Matériologiques, Paris.

Dufourmentelle, A. (2011). *Éloge du risque*. Payot, Paris.

Durkheim, E. (1970). L'individualisme et les intellectuels. In *La science sociale et l'action*. PUF, Paris.

Einstein, A. (1991). *Œuvres choisies*. Le Seuil/CNRS, Paris.

Enriquez, E. (2017). Groupe, leader bienveillant et créativité. In *La créativité au travail*, Amado, G., Bouilloud, J.P., Lhuilier, D., Ulmann, A.L. (eds). Erès, Paris.

Eklund, A.R., Navarro Aguiar, U., Amacker, A. (2022). Design thinking as sensemaking – Developing a pragmatist theory of practice to (re)introduce sensibility. *Journal of Product Innovation Management*, 39, 24–43.

Fang, Z., Song, H., Zhang, Y., Jin, B., Wu, J., Zhao, Q., Xie, T. (2020). Modular 4D printing via interfacial welding of digital light-controllable dynamic covalent polymer networks. *Matter*, 2, 1–11.

Fox, S. (2019). Mass imagineering, mass customization and mass production: Complementary cultures for creativity, choice and convenience. *Journal of Consumer Culture*, 19, 67–81.

Friedmann, G. (1963). *Où va le travail humain ?* Gallimard, Paris.

Gaglio, G. (2011). Une innovation, de nouvelles pratiques sociales. In *Sociologie de l'innovation*, Gaglio, G. (ed.). PUF, Paris.

Ge, Q., Sakhaei, A.H., Lee, H., Dunn, C.K., Fang, N.X., Dunn, M.L. (2016). Multi-material 4D printing with tailorable shape memory polymers. *Scientific Reports*, 6, 31110.

Go, K. and Carroll, J.M. (2004). The blind men and the elephant: Views of scenario-based system design. *Interactions*, 11, 44–53.

Godin, B. (2015). Innovation: A conceptual history of an anonymous concept [Online]. Available at: http://www.csiic.ca/PDF/WorkingPaper21.pdf.

von Goethe, J.W. (2004). *Faust*. Le livre de Poche, Paris.

Goetzmann, A. (2021). 8 façons de ne pas perdre confiance [Online]. Available at: https://www.contrepoints.org/2021/12/05/415739-8-facons-de-ne-pas-perdre-confiance.

Goo, B., Hong, C.H., Park, K. (2020). 4D printing using anisotropic thermal deformation of 3D-printed thermoplastic parts. *Materials & Design*, 188, 108485.

Goodhart, D. (2020). *La tête, le cœur et la main – La lutte pour la dignité et le statut social au XXIe siècle*. Les Arènes, Paris.

Goux-Baudiment, F. (2002). *Une nouvelle étape de la prospective : la prospective opérationnelle*. Pontifical Gregorian University, Rome.

Graeff, E. (2020). Innovation bio-inspirée : modélisation d'un processus interdisciplinaire de conception biomimétique outillé et intégration d'un nouvel acteur, le Biomiméticien. Thesis, ENSAM, Paris [Online]. Available at: https://tel.archives-ouvertes.fr/tel-02951019/document.

Gregor, S. and Hevner, A.R. (2013). Positioning and presenting design science research for maximum impact. *MIS Quarterly*, 37, 337–356.

Grenier, J.-Y. and Lepetit, B. (1989). L'expérience historique – À propos de C.-E. Labrousse. *Annales ESC*, 6, 1337–1360.

Guilford, J.P. (1956). The structure of intellect. *Psychological Bulletin*, 53, 267–293.

Guillin, V. (2015). Scientisme et politique. In *Sciences, technologis et sociétés de A à Z*, Prud'homme, J., Doray, P., Bouchard, F. (eds). Presses de l'Université de Montréal, Montreal.

Haleem, A., Javaid, M., Singh, R.P., Suman, R. (2021). Significant roles of 4D printing using smart materials in the field of manufacturing. *Advanced Industrial and Engineering Polymer Research*, 4, 301–311.

Hall, J. and Watson, W.H. (1970). The effects of a normative intervention on group decision-making performance. *Human Relations*, 23, 299–317.

Hamel, J. (2002). La pédagogie comme pivot de l'interdisciplinarité. *Revue internationale d'éducation de Sèvres*, 30, 143–151.

Haven, T., Pasman, H.R., Widdershoven, G., Bouter, L., Tijdink, J. (2020). Researchers' perceptions of a responsible research climate: A multi focus group study. *Science and Engineering Ethics*, 26, 3017–3036.

von Hayek, F. (1953). *Scientisme et sciences sociales*. Plon, Paris.

Hevner, A.R., March, S.T., Park, J., Ram, S. (2004). Design science in information systems research. *MIS Quarterly*, 28, 75–106.

Himanen, P. (2001). *L'éthique hacker*. Exils, Paris.

Hohberger, J. (2016). Diffusion of science-based inventions. *Technological Forecasting and Social Change*, 104, 66–77.

Hollan, J., Hutchins, E., Kirsh, D. (2000). Distributed cognition: Toward a new foundation of human-computer interaction research. *Transactions on Computer-Human Interaction*, 7, 174–196.

Hottois, G. (2009). La technoscience met-elle en danger la diversité culturelle ? *Journal international de bioéthique*, 20, 17–30.

Ininou, Y. and Loilier, T. (2021). La sélection des idées incrémentales et radicales : les enseignements d'une étude de cas. *Innovations*, 66, 139–175.

Jacob, F. (1987). *La statue intérieure*. Odile Jacob, Paris.

Jeacle, I. (2017). The popular pursuit of DIY: Exploring the role of calculative technologies in an actor network. *Management Accounting Research*, 35, 99–109.

Jeong, H.Y., An, S.C., Seo, I.C., Lee, E., Ha, S., Kim, N., Jun, Y.C. (2019). 3D printing of twisting and rotational bistable structures with tuning elements. *Scientific Reports*, 9(324).

Kelly, B., Papanikolaou, D., Seru, A., Taddy, M. (2021). Measuring technological innovation over the long run. *American Economic Review: Insights*, 3, 303–20.

Kin, V., Rollet, A., Senkel, M.P., Jan, F. (2021). Génération et sélection d'idées dans le processus d'innovation : cas de la méthodologie Delphi dans la logistique 4.0. *Innovations*, 66, 109–138.

Klein, J.T. (2010). A taxonomy of interdisciplinarity. In *Oxford Handbook of Interdisciplinarity*, Frodeman, R., Klein, J.T, Mitcham, C. (eds). Oxford University Press, Oxford.

Kline, S. and Rosenberg, N. (1986). An overview of innovation. In *The Positive Sum Strategy: Harnessing Technology for Economic Growth*, Landau, R. and Rosenberg, N. (eds). National Academy of Sciences (NAS), Washington.

Koestler, A. (1964). *Le cri d'Archimède*. Calmann-Lévy, Paris.

Lacaux, C., Andrillon, T., Bastoul, C., Idir, Y., Fonteix-Galet, A., Arnulf, I., Oudiette, D. (2021). Sleep onset is a creative sweet spot. *Science Advances*, 7(50) [Online]. Available at: https://www.science.org/doi/10.1126/sciadv.abj5866.

Landrain, T., Mayer, M.A., Perez, M., Sussan, R. (2013). Do-it-yourself biology: Challenges and promises for an open science and technology movement. *Systems and Synthetic Biology*, 7, 115–126.

Latour, B. (1984). *Les Microbes – Guerre et paix*. Métailié, Paris.

Lavis, A. (2021). *L'Imprévu – Que faire lorsqu'on ne sait plus ?* Autrement, Paris.

Le Bas, C. (2016). Les inventeurs prolifiques comme déterminants des performances technologiques des firmes. *Innovations*, 50, 119–141.

Le Crosnier, H., Neubauer, C., Storup, B. (2013). Sciences participatives ou ingénierie sociale : quand amateurs et chercheurs co-produisent les savoirs. In *Interdisciplinarité : entre disciplines et indiscipline*. CNRS, Paris.

Le Guyader, H. (2008). Morphogenèse, structures physiques et évolution biologique. *Comptes Rendus Chimie*, 11, 186–191.

Lechevalier, E. (2019). *Innovation Beyond Technology: Science for Society and Interdisciplinary Approaches*. Springer, Singapore.

Lecouvey, G., Gonneaud, J., Eustache, F., Desgranges, B. (2015). Les grandes théories de la mémoire prospective : vers une vision dynamique des processus cognitifs engagés lors du rappel programmé d'intentions. *Revue de neuropsychologie*, 7, 207–216.

Lepage, H. (2015). *Libéralisme et liberté d'expression*. Texquis, Paris.

Levin, J. (2010). *Un fou rêve de machine de Turing*. Markus Haller, Geneva.

Lévi-Strauss, C. (1990). *La pensée sauvage*. Pockett, Paris.

Lévy-Leblond. J.M. (2015). Il n'y a pas de maîtrise démocratique de la science [Online]. Available at: https://sciences-critiques.fr/jean-marc-levy-leblond-le-developpement-de-la-science-nest-pas-democratique/.

Libmann, F., Breesé, P., Versailles, D.W., Mérindol, V. (2011). Innover, oui mais comment ? *Documentaliste-Sciences de l'Information*, 48, 38–49.

Listek, V. (2019). Interview with Nicole Hone who uses 4D printing to make tangible animation [Online]. Available at: https://3dprint.com/237494/interview-with-nicole-hone-who-uses-4d-printing-to-make-tangible-animation/.

MADEELI (2017). Fabrication additive : enjeux et impacts en Occitanie/Pyrénées-Méditerranée [Online]. Available at: http://www.madeeli.fr/wp-content/uploads/2017/01/Fabrication-Additive-en-Occitanie-Etude-2016.pdf.

Mast, J., Ahrweiler, P., Viale, R. (2013). Cultural theory and its spaces for invention and innovation. *Mind & Society*, 12, 23–33.

Mauss, M. (1950). *Sociologie et anthropologie*. PUF, Paris.

Mayor, F. (1992). *Entre savoirs : l'interdisciplinarité en acte ; enjeux, obstacles, résultats*. Erès, Paris.

McDougall, J. (1978). *Plaidoyer pour une certaine anormalité*. Gallimard, Paris.

Melchels, F.P.W., Domingos, M.A.N., Klein, T.J., Maldo, J., Bartolo, P.J., Hutmacher, D.W. (2012). Additive manufacturing of tissues and organs. *Progress in Polymer Science*, 37, 1079–1084.

Mellor, S., Hao, L., Zhang, D. (2014). Additive manufacturing: A framework for implementation. *International Journal of Production Economy*, 149, 194–201.

Mitcham, C. (1998). The importance of philosophy to engineering. *Teorema*, 17, 27–47.

Mizeret, J. (2015). Les technologies de fabrication additive pour la créativité, le prototypage, la fabrication [Online]. Available at: http://www.swissmem.ch/fr/presentations-fabrication-additive.html.

Monod, J. (2014). *Le hasard et la nécessité – Essai sur la philosophie naturelle de la biologie moderne*. Points Essais, Paris.

Moreno, R. (1990). *Théorie du bordel ambiant – Souvenirs de l'irréversible*. Belfond, Paris.

Naudon, F. (2013). Comment le profane joue en faveur du décloisonnement. In *Interdisciplinarité : entre disciplines et indiscipline*. CNRS, Paris.

NCSE (2021). Understanding unsuccessful innovation. Working document [Online]. Available at: https://www.nsf.gov/statistics/2021/ncses21201/ncses21201.pdf.

Nelson, R. and Winter, S. (1982). *An Evolutionary Theory of Economic Change*. Holland Press, Amsterdam.

Nicolescu, B. (1996). *La transdisciplinarité ; manifeste*. Éditions du Rocher, Monaco.

Nietzsche, F. (1982). *The Cay Science*. Gallimard, Paris.

Nowotny, H. (2015). The radical openness of science and innovation: Why uncertainty is inherent in the openness towards the future. *EMBO Reports*, 16, 1601–1604.

O'Neal, B. (2020). International research yields new concepts for control-based 4D printing in adaptive systems [Online]. Available at: https://3dprint.com/266678/international-research-yields-new-concepts-for-control-based-4d-printing-adaptive-systems/.

Ogilvie, S. (2019). *The European Guilds: An Economic Analysis*. Princeton University Press, Princeton.

Olle-Vila, A., Duran-Nebreda, S., Conde-Pueyo, N., Montanez, R., Sole, R. (2016). A morphospace for synthetic organs and organoids: The possible and the actual. *Integrative Biology*, 8, 485–503.

Palmberg, C., Pajarinen, M., Nikulainen, T. (2010). Transferring science-based technologies to industry – Does nanotechnology make a difference? *The Research Institute of the Finnish Economy* [Online]. Available at: https://www.researchgate.net/publication/24123251_Transferring_Science-based_Technologies_to_Industry_-_Does_Nanotechnology_Make_a_Difference.

Pareto, V. (1965). *Ecrits sur la courbe de la distribution de la richesse - Œuvres complètes*. Droz Ed, Paris.

Paris, T. (2010). *Manager la créativité. Innover en s'inspirant de Pixar, Ducasse, les Ateliers Jean Nouvel, Hermès…* . Pearson, Paris.

Parmentier, G., Paris, T., Gandia, R. (2021). De la créativité à l'innovation : paradoxes, articulation et enjeux. *Innovations*, 66, 5–16.

Pascal, A. (2012). Le design science dans le domaine des systèmes d'information : mise en débat et perspectives. *Systèmes d'information et management*, 17, 7–31.

Pascal, A. and Rouby, E. (2017). Les scénarios d'usage comme support aux méthodologies de recherche en design science dans le cas d'invention. *Systèmes d'information et management*, 22, 47–78.

Peffers, K., Tuunanen, T., Rothenberger, M.A., Chatterjee S. (2007). A design science research methodology for information systems research. *Journal of Management Information Systems*, 24, 45–77.

Peng, W., Unutmaz, D., Ozbolat, I.T. (2016). Bioprinting towards physiologically relevant tissue models for pharmaceutics. *Trends in Biotechnology*, 34, 722–732.

Philippe, G. (2021). Qu'est-ce que la fabrication additive ? [Online]. Available at: https://www.3dnatives.com/fabrication-additive/.

Piaget, J. (1967). *Logique et connaissance scientifique*. Gallimard, Paris.

Pineo, H., Turnbull, E.R., Davies, M., Rowson, M., Hayward, A.C., Hart, G., Johnson, A.M., Aldridge, R.W. (2021). A new transdisciplinary research model to investigate and improve the health of the public. *Health Promotion International*, 36, 481–492.

Pingeot, M. (2022). Cette quête éternelle de l'immortalité [Online]. Available at: https://up-magazine.info/en/decryptages/99344-cette-quete-eternelle-de-limmortalite/.

Pinkerton, A.J. (2016). Lasers in additive manufacturing. *Optics & Laser Technology*, 78, 25–32.

Popper, K. (1989). *The Logic of Scientific Discovery*. Payot, Paris.

Portella, E. (1992). *Entre savoirs : l'interdisciplinarité en acte ; enjeux, obstacles, résultats*. Erès, Paris.

Protolabs (2020). L'ABC du prototypage [Online]. Available at: https://www.protolabs.fr/ressources/livres-blancs/l-abc-du-prototypage/.

Rancière, J. (1994). Histoire des mots, mots de l'histoire. *Communications – L'écriture des sciences de l'homme*, 58, 87–101.

Raynaud, D. (2021). *Sociologie fondamentale – Étude d'épistémologie*. Éditions Matériologiques, Paris.

Richards, D., Jia, J., Yost, M., Markwald, R., Mei, Y. (2017). 3D bioprinting for vascularized tissue fabrication. *Annals of Biomedical Engineering*, 45, 132–147.

Roberts, E.W. (2007). Managing invention and innovation: Technological innovation can alter the competitive status of firms and nations but its purposeful management is complex, involving the effective integration of people, organizational processes and plans. *Research Technology Management*, 50, 35–54.

Roco, M. and Bainbridge, W.S. (2013). The new world of discovery, invention, and innovation: Convergence of knowledge, technology, and society. *Journal of Nanoparticle Research: An Interdisciplinary Forum for Nanoscale Science and Technology*, 15, 1–17.

Rodriguez, D. (2021). L'histoire de la Fabrication Additive. Partie I [Online]. Available at: https://tessellated.eu/fr/blog/histoire-fabrication-additive-partie-i/.

Rodrik, D. (2009). *Growth After the Crisis, Mimeo*. Harvard Kennedy School, Boston.

Romian, H. (2000). Pour une culture commune : de la maternelle à l'université. In *Des problématiques complexes – Transformer l'école et la culture scolaire*, Romian, H. (ed.). Institut de recherches de la FSU Hachette Éducation, Paris.

Rouquette, M.L. (2007). L'invention de l'invention. In *La créativité*, Rouquette, M.L (ed.). PUF, Paris.

Rouquette, M.L. (2016). Ingénierie de la création. In *Innovation sur mesure*. De Boeck Supérieur, Paris.

Rousseau, J.J. (2009). *L'Émile*. Garnier-Flammarion, Paris.

Sarooghi, H., Libaers, D., Burkemper, A. (2015). Examining the relationship between creativity and innovation: A meta-analysis of organizational, cultural, and environmental factors. *Journal of Business Venturing*, 30, 714–731.

Sarpong, D., Ofosu, G., Botchie, D., Clear, F. (2020). Do-it-yourself (DIY) science: The proliferation, relevance and concerns. *Technological Forecasting & Social Change*, 158, 120127.

Schumpeter, J.A. (1942). *Capitalism, Socialism and Democracy*. Harper & Row, New York.

Schwartz, J.J. and Boydston, A.J. (2019). Multi-material actinic spatial control 3D and 4D printing. *Nature Communications*, 10(791).

Sein, M.K., Henfridsson, O., Purao, S., Rossi, M., Lindgren, R. (2011). Action design research. *MIS Quarterly*, 35, 37–56.

Serres, M. (2013). Non, le réel n'est pas découpé en créneaux. In *Interdisciplinarité : entre disciplines et indiscipline*. CNRS, Paris.

Shaw, M.E. (1932). A comparison of individuals and small groups in the rational solution of complex problems. *American Journal of Psychology*, 44, 491–504.

Shen, Z.C., Xia, Y., Ding, Y.G., Zhao, C.Q., Yang, Y.B. (2019). 4D printing and its key technologies. *Journal of Materials Engineering*, 47, 11–18.

Silberstein, M. (ed.). (2013). Conclusion. Science(s) et matérialisme(s) : examen des conditions d'une synonymie. In *Matériaux philosophiques et scientifiques pour un matérialisme contemporain. Volume 2 – Sciences, ontologie, épistémologie*. Éditions Matériologiques, Paris.

Silberzahn, P. (2013). Lean startup : le compagnon idéal de l'effectuation ? *Entreprendre & Innover*, 19, 29–35.

Sommer, P. (2016). Les tissus humains, entre biologie et mécanique – Propos d'un biologiste. In *MECAMAT – CNRS*. Aussois.

Sorenson, O. and Fleming, L. (2004). Science and the diffusion of knowledge. *Research Policy*, 33, 1615–1634.

Stahl, G. (2011). Theories of cognition in CSCW. In *Proceedings of the 12th European Conference on Computer Supported Cooperative Work Ed.* Aarhus [Online]. Available at: https://www.researchgate.net/publication/220265933_Theories_of_cognition_in_CSCW.

Stengers, I. (2014). Ralentir les sciences, c'est réveiller le chercheur somnambule. *Écologie & politique*, 48, 61–74.

Stokes, D.E. (1997). *Pasteur's Quadrant: Basic Science and Technological Innovation.* Brookings Institution, New York.

Taleb, N.N. (2011). *Le Cygne noir, la puissance de l'imprévisible.* Les Belles Lettres, Paris.

Tang, Y., Qin, L., Li, X., Chew, C., Zhu, J. (2017). A frog-inspired swimming robot based on dielectric elastomer actuators. In *IEEE/RSJ International Conference on Intelligent Robots and Systems (IROS)*. Vancouver.

Tarde, G. (1902). L'invention, moteur de l'évolution sociale. *Revue internationale de sociologie*, 10, 562–574.

Theys, J. and Cornu, P. (2019). Temps de la nature, temps de la société, temps scientifique à l'heure du changement global. Pour une approche interdisciplinaire de la crise de la temporalité. *Natures sciences sociétés*, 27, 381–389.

Thomas, R., Basiosa, V., Eiswirth, M., Kruel, T., Rossler, O.E. (2004). Hyperchaos of arbitrary order generated by a single feedback circuit, and the emergence of chaotic walks. *Chaos: An Interdisciplinary Journal of Nonlinear Science*, 14(699).

Thomas, C., Tibbits, S., Banning, G. (2014). The programmable world. *Scientific American*, 311, 60–65.

Tibbits, S. (2014). 4D printing: Multi-material shape change. *Architectural Design, Special Issue: High Definition: Zero Tolerance in Design and Production*, 84, 116–121.

Tibbits, S. (2016). *Self-Assembly Lab: Experiments in Programming Matter.* Routledge, New York.

Tibbits, S. and Tomas, A.F. (2013). Biomolecular, chiral and irregular self-assemblies. *Acadia Adaptative Architecture*, 261–268 [Online]. Available at: http://papers.cumincad.org/data/works/att/acadia13_261.content.pdf.

Tinland, F. (1994). Le site de la technique : éclairages théoriques et enjeux pratiques. In *Ordre biologique ; ordre technologique*, Tinland, F. (ed.). Champ-Vallon, Paris.

Tranchard, S. and Rojas, V. (2015). L'avenir en fabrication 3D [Online]. Available at: http://www.iso.org/iso/fr/home/news_index/news_archive/news.htm?refid=Ref1956.

Valéry, P. (1957). *Œuvres*. Gallimard, Paris.

Valke, P. (2021). Pour en finir avec les disciplines ? [Online]. Available at: https://themetanews.startup-guyane.tech/index.php/2021/01/.

Vandeginste, P. (2012). Modéliser le vivant [Online]. Available at: http://www.larecherche.fr/savoirs/dossier/4-modeliser-vivant-01-05-2012-91189.

Viénot, J.M. (2022). Comment la mode s'empare de l'impression 3D pour créer les vêtements et accessoires de demain [Online]. Available at: https://www.bfmtv.com/tech/com-ment-la-mode-s-empare-de-l-impression-3d-pour-creer-les-vetements-et-accessoires-de-de-main_AN-202204180027.html.

Vinck, D. (2007). *Sciences et société ; sociologie du travail scientifique*. Armand Colin, Paris.

Wang, Q., Yang, Z., Lu, Z., Li, X. (2020). Mechanical responses of 3D cross-chiral auxetic materials under uniaxial compression. *Materials & Design*, 186, 108226.

Weller, C., Kleer, R., Piller, F. (2015). Economic implications of 3D printing: Market structure models in light of additive manufacturing revisited. *International Journal of Production Economics*, 164, 43–56.

Wiener, N. (1997). *L'utopie de la communication*. La Découverte, Paris.

Wilson, R.C., Guilford, J.P., Christensen, P.R. (1953). The measurement of individual differences in originality. *Psychological Bulletin*, 50, 362–370.

Wittrock, C., Forsberg, E.M., Pols, A., McNaghten, P., Ludwig, D. (2021). *Implementing Responsible Research and Innovation: Organizational and National Conditions*. Springer International Publishing, Cham.

Wolton, D. (2013). Conclusion – Pour un manifeste de l'indiscipline. In *Interdisciplinarité : entre disciplines et indiscipline*. CNRS, Paris.

Yao, Y., Waters, J.T., Shneidman, A.V., Cui, J., Wang, X., Mandsberg, N.K., Li, S., Balazs, A.C., Aizenberg, J. (2018). Multi-responsive polymeric microstructures with encoded predetermined and self-regulated deformability. *PNAS*, 115, 12950–12955.

Zolfagharian, A., Kaynak, A., Kouzani, A. (2020). Closed-loop 4D-printed soft robots. *Materials and Design*, 188, 108411.

3

Creativity and Additive Manufacturing

The fundamental discoveries that are still the basis of our existence: agriculture, animal husbandry, pottery, weaving, were the work of very primitive societies. Could it be said that they were the fruit of chance? By analyzing the main techniques, I aim to show the naivety of such an interpretation: "All these operations are far too numerous and complex for chance to account for. Each of them, taken in isolation, means nothing, and it is their combination, imagined, desired, researched and experimented with, that alone enables success. Chance may exist, but it doesn't in itself produce results. For some two and a half thousand years, the Western world has known about the existence of electricity – discovered no doubt by chance – but this chance was to remain sterile until the intentional, hypothesis-driven efforts of the Ampères and Faradays. Chance played no greater role in the invention of the bow of the bow, the boomerang or the blowpipe, in the birth of agriculture and animal husbandry, than in the discovery of penicillin – from which, incidentally, it was not absent". The real problem lies elsewhere: "namely, that, despite a dose of imagination, invention and creative effort, which we have every reason to suppose remains more or less constant throughout the history of mankind, this combination determines major cultural mutations only at certain periods and in certain places. For, to achieve this result, purely psychological factors are not enough: they must first be present, with a similar orientation, in a sufficient number of individuals for the creator to be immediately assured of an audience; and this condition itself depends on the coming together of a considerable number of other factors, of a historical, economic and sociological nature".

(Lévi-Strauss 2020)

For a color version of all the figures in this chapter, see www.iste.co.uk/andre/knowledge2.zip.

3.1. Toward a Big Bang of creativity

> Theory is when you know everything and nothing works; practice is when everything works and nobody knows why. Here we combine theory with practice: nothing works and nobody knows why (Einstein 2007).

> Results are achieved by exploiting opportunities, not by solving problems (Drucker 2006)[1].

> A search is first and foremost a journey, which brings it closer to error, which we have seen is also based on a journey. The notion of a journey itself presupposes that we move from one place to another with the intention of achieving a goal. In the case of research, however, several obstacles stand in the way. The first is ignorance of the wife. If it were already known, there would be no more research. But it's not even a clear-cut path, otherwise you wouldn't get off the beaten track. So we have to make our own way, which involves a form of wandering (Lambert 2015).

EPSTs (public scientific and technical establishments in France) and universities praise the merits of "interdisciplinary", "multidisciplinary" approaches, innovation, etc., in their activity reports or in their programmatic documents. The reference to interdisciplinarity alone is an ambiguous reference to an indeterminate social object but, in fact, it is a functional reference (Hatano-Chalvidan 2010): it is not a goal, but a means, for example, used for invention or innovation. In these last fields, an epistemological reversal must be taken into account: scientific and technical culture no longer determines action; it is action that mobilizes this culture.

In any research, there is always a risk of error, which means that we are not taking the same routes as others. The existence of standardized methodologies limits this situation. And yet, in principle, it is the unknown that should drive research. Christopher Columbus would not be too far off... In reality, most research is incremental in nature, which means that risk-taking is fairly well controlled. With disruptions (inventions, concepts, etc.), the only safeguards for maintaining rationality seem to lie in the culture of disruptive researchers with a responsible freedom. This section considers extensions to this assertion and shows that between the basic idea that Lavelle (1992) calls "confused apprehension" and the validation of the idea in terms of credibility (other stages will be necessary before it becomes an invention), various contexts need to be clarified, particularly in the immediate

1 Author's translation from the French version of the text.

vicinity of the emergence of this "confused apprehension", hence the title of section 3.1.

3.1.1. *Background and theoretical considerations*

> When the cleaning was done, he [...] turned it upside down and broke it with several hammer blows, then picked up the scattered fragments. In this way, the pot took up very little space and would fit into a standard model box, without detracting from the collections (Vian 1956)[2].

> Its 'fuel' is dopamine, the neurotransmitter of positive emotions. This chemical messenger directs our behaviors and choices according to their contribution to the ultimate Darwinian goal: the dissemination of our genes through survival and fertility (Cropley 2015)[3].

> The power of metaphor comes from its ability to draw our attention to the most salient challenges so as to anticipate and, ideally, overcome them (Boffa-Comby 2019).

> Heuristics are efficient cognitive processes, conscious or unconscious, that ignore part of the information (Gigerenzer and Gaissmaier 2011).

What honest, principaled man, who is subjected daily to propagandistic talk about the governments of science and invention, can look with a priori benevolent and sometimes worried curiosity at the evolution of these fields, considered as major, as essential to the survival of the human species? If there is no idealization of earlier periods when the world was to be discovered with just our heads and a few instruments, because we live in the present, this double knowledge of education (historical knowledge) and daily observation enables us to analyze the emergence of ideas (whether profitable or not), which contrasts with the enthusiasm and overuse of concepts (or their mere display) such as interdisciplinarity, demand-driven management, "radical" methodologies to support creativity, etc.

The observations presented below are interspersed with more personal reflections, on his fear of a display on creation freeing itself from its foundations, with its difficulties and modest successes. Obviously, the comments made in this chapter will be used to analyze the positioning of additive manufacturing technologies, 4D printing and bio-printing.

---

2 Author's translation from the French version of the text.

3 Author's translation from the French version of the text.

3.1.2. *Introduction*

> To research is to invent the world, to set new rules for an ephemeral world. Not like tyrants, who invent another world for themselves but impose it on others. The researcher doesn't remake the world; he undoes it in order to make it. He imagines one, then confronts it with the real world to illuminate it, not exhaust it. Seeking is a never-ending quest (Rose 2001).

> At first glance, moral theory and ethics seem to be preoccupied with the same question: 'What must I do?'; 'What must we do?'. Only 'ought' no longer has the same meaning when we ask ourselves about the rights and duties we impute to everyone reciprocally from the point of view of the first person plural, and when, concerned about our own lives, we adopt the point of view of the first person singular to ask ourselves what, all things considered, is better to do in the long run 'for me' or 'for us' (Habermas 2002).

> White-collar activism is bad, enthusiasm doesn't starch (Elgozy 1966).

Do you remember the distant days of your last prize-giving ceremony? You waited, a little nervous, your hands slightly sweaty, under your grandfather's watchful eye, for the principal to mention your name for the prize for camaraderie, mathematics or best chalkboard eraser. You would climb up to the podium, intimidated on the way out, careful not to miss a step, triumphant-eyed on the way back, delighted with a book that would soon be forgotten and a handshake from the holder of power, sure that you were on the path mapped out by pedagogues and school curricula. We (they?) had been shaped/formatted in the image of the value system of the day, and our future authority over the world was magnified by these inaugural acts! A kind of pedagogical communion of secular initiation! All of those who had been honored were holders of knowledge, in the image of our masters, priests initiating knowledge that was generally scientific, robust, established, reassuring and therefore right.

Within this framework, initial training is usually purely scientific, combined with uncritical submission to the "dogmas" handed down by our educators (we all have to pass our exams). So how do we move from the known to the imagined? From conformity to invention? From rigid but efficient training to dynamic research and/or transgressions linked to inventions? There are many answers to these questions, which lie well outside the realm of "simple" scientific research. They stem from the image of the researcher as a judge of human activity, always the holder of (good) knowledge, from the desire to pursue our dreams through imagination and creativity, or from the need for recognition, etc.

François Kourilsky, former Director General of the CNRS, spoke on the subject of interdisciplinarity and considered that the issue of the difficult development of science and innovation is not focused on this "constructive indiscipline" but on the disciplines themselves. In 2002, he wrote:

> How many researchers push specialization to the point of disinterest in crossing the boundaries of their discipline. Worse still, some are no longer even interested in the frontiers of the discipline itself.

A number of questions emerge from these remarks: are there "naturally" creative people? How does the research system deal with small, idea-provoking disruptions? Is it easy to accept this form of epistemological anarchism (Feyerabend 1988)? Other questions are raised in this chapter.

Some envisioned a future that would be at least partially programmable, because it would be possible to exploit conceptual knowledge created within a framework regulated by financially supported calls for tender with financial backing, a far cry from the responsible independence that every "old" researcher has had the good fortune (?) not to practice. The effect of the creation of funding agencies may correspond (at least for the author) to a transgression of previous rules (which, incidentally, had no reason to be particularly perpetuated in the space of changing values in which we live (Grieu 2006). These new rules envisage the development of axes that are favored (by whom?) because they respond to a need that can be expressed in the short, or perhaps even medium term. The result of this new situation is more targeted funding, combined with the attraction of the best brains for priority areas, with a view to the best possible operational efficiency.

However, the expected long-term effect is twofold: creativity potentially diminished, but greater "disciplinary professionalism", with the sidelining of certain scientific fields, or even a loss of initiative. This new situation does, however, lead to a more "Darwinian" life for researchers (Drucker 2006), with those who are best able to cope in the emerging normative universe surviving, thanks to changes in the "institutional" research environment, and a good perception of the new explicit and, above all, implicit rules of operation. Specialization is probably a necessary evil in the sciences for reasons of efficiency. But how can we avoid premature compartmentalization, or even unsuitable programs that risk "ossifying" specialists, walled in by their own systems of thought (Elgozy 1966)? Ganascia (1999) writes that "specialization gradually deforms the mind to the point of crippling it [...]". Each individual, in their isolation and narrow-mindedness, now grasps only the crumbs of a whole in its obvious totality. So, how do we come up with the right idea, one that can contribute to scientific and/or technological resources?

Figure 3.1, taken from Le Comptoir (2018), reminds us that technological evolution works in leaps and bounds, followed by improvements. One of the questions to be addressed in this chapter is how these disruptions came about. Let us recall that scientific disruptions can be seen as discoveries that transform the frontier of knowledge and can have an impact on science, technology and society.

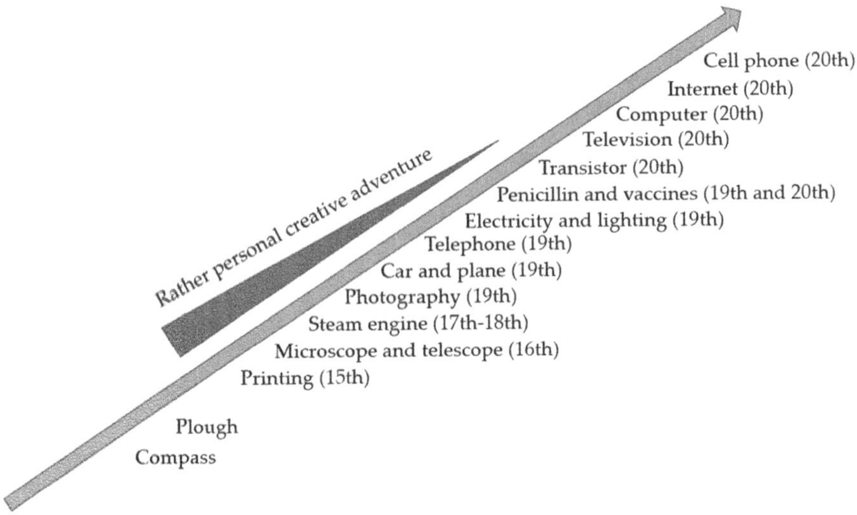

Figure 3.1. *Historical approach to disruptions and inventions*

3.1.3. *Just before the idea*

> Contemporary societies are constantly changing, with an uncontrollable acceleration: they are flexible societies, without external borders, without internal limits, fluid societies. These conditions have profound consequences for the way we behave (Haroche 2006).

> Emergence of an idea: this is a quantitative expansion and extension of fields of investigation [...] to previously ignored Fareas (Ansart 1990).

> You have an idea, you fight for it, you hammer it into people's heads, you watch it grow and triumph... (Zola 1997).

Thanks to this new representation of the 'world', we discover, among other things, experiential knowledge. Reflection on this new representation then leads us to better identify, articulate and share our own experiential knowledge. To better understand the process by which we construct the knowledge that comes from our practice, we need to go back to ontologically primary experience, to a point prior

> to the distinction between intuition and intelligence. In particular, this is why we speak, let's remember, of an experiential antecedent enabling the triggering of the construction of a new representation of the world, by way of a form of redefinition of the self, as a participant in this world (Gambier 2019).

Obviously, the very first steps of these possible changes induced by this trigger, that is, the idea, are particularly important, since they condition the emergence of a credible proposal. In fact, the idea emerges during the action as a contingent intrusion, because it breaks or brings new elements that come out of previously accepted knowledge (idea of a "confused apprehension"), according to Lavelle (1992), a kind of transformative learning (in a more or less radical way). It is subsequently defined by a change in representation. It is moreover from this space of prior knowledge, associated with an (old) representation, that an idea emerges.

Gambier (2019) considers that a "translation" of the idea involves a kind of hub, an exchanger where various disciplines and/or trades intersect, but without knowing exactly the conditions and effects of these crossings. To borrow from Augé (1992), the disciplines involved are more likely to be in a relationship of "consumption", of "digestion" before any reciprocity emerges, despite an initial willingness to share in/for an encounter that should be fruitful. Indeed, the notions of representation, and especially that of changing of representation, are central to making an idea credible.

For Lavelle (1992), there is a kind of sensation of tipping over into another world to propose a new representation through the idea.

> To better understand the process of constructing our knowledge from our practice, we would thus have to go back to the ontologically primary experience where we are at a point prior to the distinction between intuition and intelligence (Gambier 2019).

This is an experiential antecedent enabling the construction of a disruption to be triggered. An important point to remember is that putting forward an idea requires us to be able to think outside the box and to accept alternative proposals to the initial ones, which are themselves undeveloped. For many, the idea came about by chance! But for Mishima (1989):

> To speak of chance is to negate the possibility of any law of cause and effect. Chance is the one final irrationality acceptable to the free will. Without the concept of chance, you see, the Western philosophy of free will could never have arisen. Chance is the crucial refuge of the will. And without it the very thought of gambling would be inconceivable, just as the Westerner has no other way of rationalizing the repeated setbacks and frustrations that he must endure. I think that

> this concept of chance, of a gamble, is the very substance of the God of Europeans, and so they have a deity whose characteristics are derived from that refuge so vital to free will, namely chance – the only sort of God who would inspire the freedom of human will.

Not all inventions are of the same kind: some correspond to well-defined, single-answer problems, and others (see right-hand side of Figure 3.2) to more complex, ill-defined, open-ended problems (involving notable efforts and well-structured support groups). As Horenstein (2002) explains: "If there is only one answer to the problem, and it's simply a matter of putting the pieces together, then the activity is probably analysis[4]". This is what the "tinkerer" does, putting together scientific and technical elements, to keep an idea of the whole (exploiting an ability to correlate many ideas and information coming from different fields). The slightly skilled tinkerer thus possesses a certain culture, elements of which are organized "until suddenly – almost surprisingly – it makes sense and the innovator can say 'Eureka!" (Millier 2016). On the other hand, in vaguer and more divergent situations, the characteristics of the idea are not completely determined, and the number of solutions and/or solution paths capable of satisfying the underlying need may be high. The criteria for recognizing a solution may be open-ended (Kozbelt et al. 2010; Copley 2015). Figure 3.2 gives an overview of these different characteristics and the possible need for clarification.

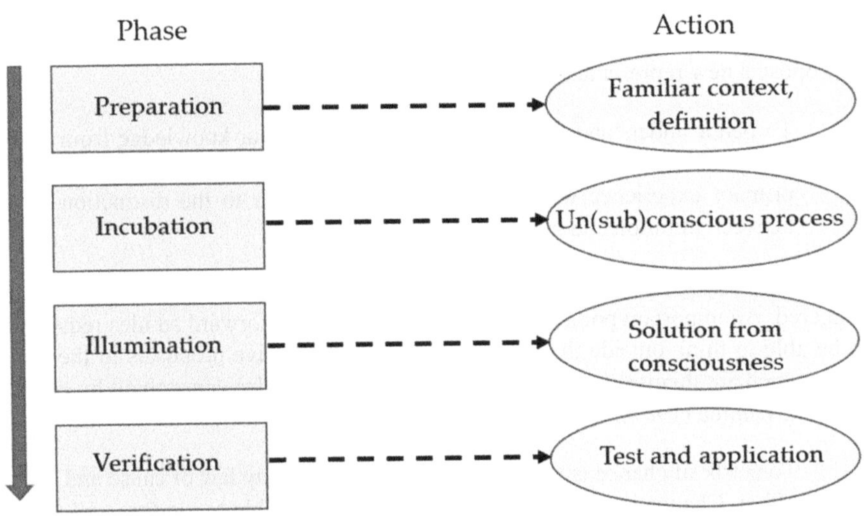

Figure 3.2. *Elements likely to be involved in the initial phase of creativity (source: Walras 1926)*

4 Author's translation from the French version of the text.

Lévy-Leblond (2006) reminds us that "the inventor of a new idea has no reason to believe in it from the outset. The greater his culture, the more numerous are the objections that arise in his mind, the situations inextricable by this idea, the irreconcilable contradictions with accepted ideas". The event induces options that may seem surprising, and which, depending on their importance, can modify customary divisions, groups, collectives, parties and communities, along unexpected lines.

NOTE.– Creative intelligence (?).

According to Lubart (2010), intelligence refers to the cognitive abilities needed to identify and define a problem or task; to master the information required for the task (selective encoding); to exploit analogies; to combine elements to create a new idea (selective combination); to exploit divergent thinking; to evaluate ourselves; and to distance ourselves from the initial idea and explore new avenues (cognitive flexibility). Thus, creative ability can only be expressed through the individual's ability to override the judgment shared by the majority (McCauley 2012).

Whatever the case, "métis" is a form of intelligence that is always "immersed in a practice", combining "flair, wisdom, foresight, an open mind, cunning, resourcefulness, vigilance, opportunism, a variety of skills, extensive experience" (Detienne and Vernant 1974).

Figure 3.3, taken from Muller (1993), highlights what Muller calls "constellations", which are directly involved in the creation of ideas.

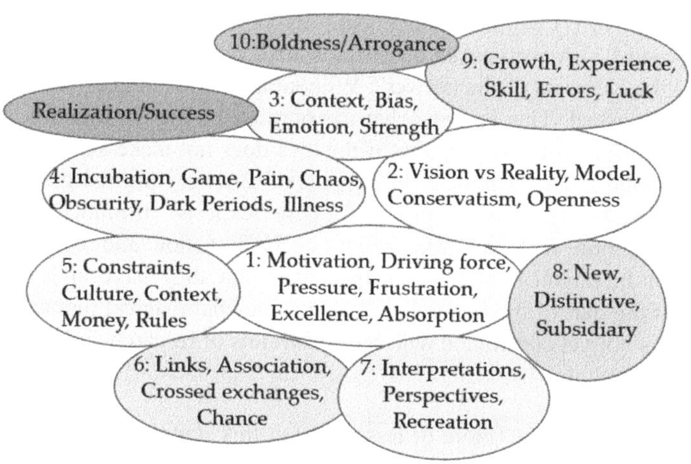

Figure 3.3. *Constellations for creativity*

Moreover, the researchers discovered that more than half of the choices usually considered to be based on curiosity were in fact due to errors in evaluation. "This result is important because it implies that many choices towards the unknown are made without our knowledge or awareness" (Findling et al. 2019)[5].

3.1.4. *Clarification*

> You can turn around a chair, seeing it from different angles, without feeling that you're dealing with several different chairs each time you switch from one vision to another. Popularization does exactly that. The variety of forms to which it appeals allows the supposedly identical content to expand with reality, to acquire relief (Jurdant 2009).

> The difficulties of transmission are often difficulties of language, the obliteration of conceptual meanings under purely formal signifiers that do not call for meaning. What, on the speaker's side, is coherently and concretely conceived is often rendered by his discourse [...] in a mutilated, desiccated way, and reduced to a mass of content, a mixture of esoteric terms and technical details, whose quantity of information [...] they carry is quite impossible to assimilate (Paty 1990).

> We are but dwarfs standing on the shoulders of giants (Saint Augustin, quoted by Lenoir (1995)).

Canguilhem (1957) wrote: "Authentically important questions are questions that are badly posed [...] A well-asked question is already no longer a question, since it contains all the elements of the answer". Without a paradox, a question can only, as such, be badly asked. This is somewhat the purpose of the clarification operation, with the first mission of verifying that if the idea does not make sense, it is deemed worthless (Cholle 2020). It is with the support of a rather informal group that does not have precise borders that we proceed. These "experts" in immediacy therefore have no connection with an organization chart in action and can easily cross conceptual or hierarchical boundaries (provided that they have at least the tacit agreement of the part of the hierarchy). It is a somewhat magical moment when the nascent idea is not yet confronted with the constraints of research and barely touches what could be a reality.

In its first moment, the (more or less robust at this stage) initiation of the idea is also ontologically prior to Archimedes' "Eureka!", that is, the moment when "we

5 Author's translation from the French version of the text.

suddenly find the solution to a problem, a good idea" (Rey 1992). Lavelle (1992) speaks of a situation he calls "confused apprehension", transition point between before and after. The creative act is a necessary condition for the production of a possible invention whose synthesis more or less intuitive, but reactive, is produced in the process. It must be elaborated and objectified to make an inventive idea accessible. "The dialectic at work during this long sequence, between the creative gesture and its full realization, thus encompasses the genesis, circulation, reception and appropriation of new knowledge, and this at a collective as well as an individual level" (Ancori 2014). Figure 3.4 gives an idea of what is expected of clarification work.

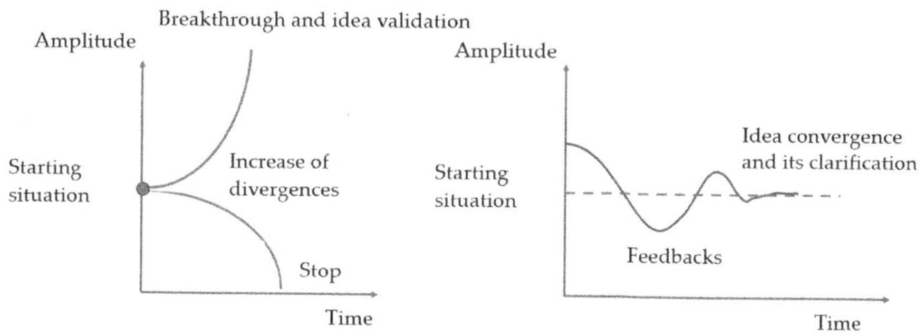

Figure 3.4. *Clarification*

> Be that as it may, the origins, the starting points of everything are not easily represented, and this is perhaps one of the dead ends or rather one of the stops of the human psyche which can only approach these beginnings in a roundabout and asymptotic way (Golse 2004).

But how would this experiential antecedent associated with the idea (experiential knowledge is knowledge resulting from practice) actually be found in the changes of representation prior to the acquisition of experiential knowledge (Gambier 2019)? The few writings on this subject concern the processes of change in these representations. The very first moments seem, for this author, particularly important in the acquisition of experiential knowledge; inevitably, they are the essential triggers for action in an act that will correspond to transformative learning: the credibility of the idea.

Initially, as part of an idea-expertise cooperation of clarification, whatever its nature, there is usually a desire to work with the other person because their practice resonates with or questions our own, to get out of the daily routine, etc. There is a desire to share, even if it means changing our point of view as a result of others' contributions. It is indeed through the renewal of our gaze and point of view that the sharing of knowledge suggests a process of "deterritorialization", to use Deleuze and Guattari's (1972) idea. By going out to meet others from other disciplines, it is possible (if we take the time) to "deterritorialize" ourselves, which not only leads to the discovery of new ways of thinking and doing, but also to the removal of our own know-how from its original context. In this way, they can be freed from constraints or alienations inherent in habitual conservatism (Bienaise 2016). Collinet et al. (2014) propose that, in order to limit interdisciplinary interactions or sequence them (above all, without eliminating them), the various scientific disciplines should reflect in isolation on the object of expertise. But this is only the first step on the staircase!

To achieve the goal of clarification, expertise must move toward a framework of solution by trying to make the positions of the various parties evolve and converge to "translate" the different languages involved. So, according to Pomerol (2012), "an expert is someone who has a good structuring of the field in mind, and who is capable of recognizing patterns he has already encountered and stored in his memory"[6]. It is all about supporting the change and stimulating the project group because it can contribute to the operation. To achieve the objective, it is necessary to understand how the complex individual/group dialectic works, converging individual energies toward the efficiency of cooperative work (Lizarralde and Pilnière 2012). Miailhe et al. (2017) remind us that we must guard against isolating parts from the whole, decontextualizing, separating external from internal factors, limiting the number of interacting dimensions, considering intercurrent events as anomalies. Yes, but this requires time and resources. They quote a Chinese proverb: "he who takes fish out of water to study their swimming produces obsolete research paradigms".

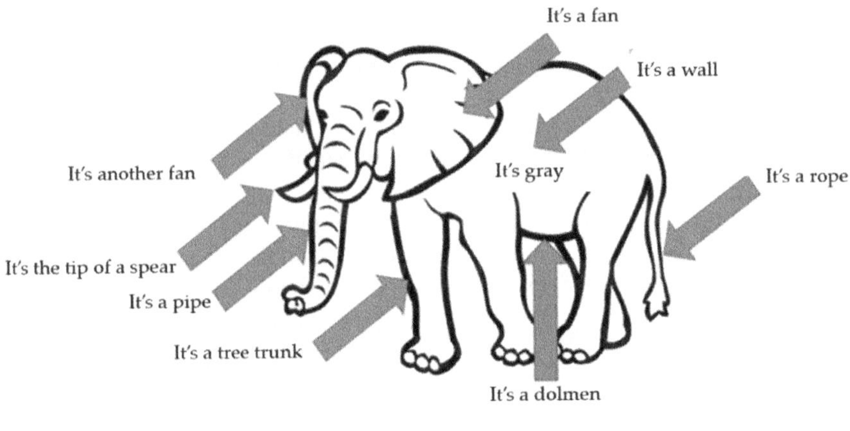

Figure 3.5. *Disciplinary difficulty in representing a complicated object*

6 Author's translation from the French version of the text.

But, to exchange validly, we must understand each other. Figure 3.5, which takes up Jurdant's (2011) remarks, clearly illustrates the need for clarification of the idea (represented in the figure by an elephant), and for this, it is necessary to call on different opinions, rather scientific, from distinct disciplines. This means that a somewhat complex idea must pass through an active filter before being accepted as a somewhat credible proposal which will then undergo other filters ("There is nothing to know if an idea is new except in reference to certain standards, nor whether it is valid until it has been socially evaluated" (Csikszentmihalyi 2006)).

Exchanges between people, apart from factual information, takes the form of argumentation. This has preoccupied thinkers since Antiquity, philosophers and jurists, rhetoricians and stylists. Indeed, to speak to one (or more) interlocutor(s) is to express emotions, to describe, to tell stories, true or false. It may involve lying, inciting, provoking, threatening, commanding, questioning, praying, promising etc. But, even in the "café du commerce", it also means trying to convince others of the truth of our proposals. A rational and, if possible, peaceful debate is not a zero-sum game, but a regulated duel of points of view, an exchange of arguments, each subject to the subtle weighing of pros and cons. The quality of the exchange of arguments is also a measure of the esteem in which the other party is held. Thus, an argument is an organized line of reasoning, an ordered set of statements, the first of which are called premises and the last, conclusion. This reasoning is valid in its presentation, if the conclusion can be derived from the premises, thanks to the use of certain rule(s) of inference. However, a rhetorically well-crafted argument can appear convincing without being valid! Sophisms and other fallacious arguments are well known, and it is possible to use logic as a weapon of criticism (robustness of deductibility). Thus, arguing often means showing your interlocutor that they cannot accept certain propositions and reject their consequences at the same time. In other words, argumentative discourse is fundamentally oriented toward a counter-discourse, in relation to which it makes sense, and whose hallmarks it often bears.

The second point concerns interdisciplinarity and its implementation. The practical barriers to setting up a multi-/interdisciplinary group cannot be underestimated. First, there are the difficulties to find a common language, to become aware of the implied, presupposed and unspoken aspects of each other's discourse, to grasp the conceptual frameworks and terminology of each discipline, which can lead to misunderstandings, and to grasp the needs and expectations of each discipline. Shared values and epistemologies in one community are revealed in ways of posing problems, selecting models, choosing variables, interpreting data, using evidence, conducting a demonstration: they are not necessarily the same in another scientific community. Resented differences can lead to defensiveness, arrogance and withdrawal into our own territory (Brownlie 2008). Unlike algorithms, the heuristics used are often drawn from experience or analogies, rather than from overly complex scientific analysis. "It reduces complexity and saves time, but the downside is that an over-simplified method can lead to cognitive biases to cognitive biases, leading to possible errors in decision-making" (Dumont 2018). Then there is the problem of choosing the right experts and the right number, given the disciplines involved (Kobarg et al. 2019).

There is indeed a moment in the deepening of the "confused apprehension" that anyone taking part in this stage of clarification approaches with a certain apprehension, that of understanding and then of interpreting the idea, because it consists of making the idea a little more intelligible without distorting it (but how do we know?), or of rejecting it. However, interpreting is not an easy task since it is necessary to understand how the proposition can lead to invention from its own representations. Moreover, we can conceive that there are several, some more goal-oriented, others more causal, even conceptual. The important point at this stage is that it is "just" a question of helping to give substance to the idea without having to justify it by in-depth actions, generally interdisciplinary. But, the practice of this clarification is more a matter of clever tinkering of an approximate crossbreeding (restorative and constructive helping relationship). Interbreeding is opposed to logos. "It is the art of cunning, of practical intelligence, of crossroads. Knowledge is reappropriated, cobbled together" (Foucart 2008).

In this context, the notion of judgment heuristics (rapid cognitive shortcuts for simplifying an analysis or decision) is highly effective in situations where speed can be of the essence (Bazerman and Moore 2009).

However,

> it is necessary to gather and structure a hierarchy of skills and expertise that are capable of collective evaluation. In doing so, one inevitably creates a group of an ambiguous nature which not only possesses collective expertise, but which has ambitions, fears, collective interests and in which individual ambitions and interests are rooted (Lebeau 1996).

Using your address book can be very useful in selecting experts. This argued synthetic activity, regardless of the experts, is not obvious (except in simple cases). While it is a slow elaboration (Dumora 2020), it depends on minds driven by curiosity, with a great sensitivity for what is new, "in the sense that the new must be described, recorded, but also interpreted, in order to integrate into the known world" (Demora 2020).

The end of the operation means that a provisional decision has been taken and therefore legitimized. This action solution presupposes an appreciation of its operational aspects, a knowledge of its parameters and, if possible, an idea of the possible outcomes by comparing the potential merits of several options. Anticipating and explanation of potential effects are part of the expected consequences.

As you can see, these activities are not self-evident, and require the partners involved to take risks concerning the future of the idea, especially as the decisions taken are part of a competitive environment that could affect it:

> What is deciding without solid foundations, in complexity and vagueness? What is it to act without pre-established trust in institutions, in the absence of stable norms? How do precaution, prudence, flexibility and reversibility become resources facilitating (and constraints limiting) decision and action? How does the establishment of various forms of deliberation help to face insoluble dilemmas? How can practical wisdom or reflexivity serve as a guiding principle for action? The list of questions is not closed. On the contrary, the field seems very broad, the problem would rather be to circumscribe its perimeter (Soulet 2017).

NOTE.– Academic and industrial worlds.

Change requires involvement, a willingness to question what needs to be questioned, and the ability to create a sense of shared responsibility. In the industry, commitment can be strongly "desired", with initiative taken on a regular basis. This desire to encourage collective success is not found in the academic world. In the production sector, the experts are those of the company. They can be recognized by their ability to look reality in the face, with objectivity and pragmatism, and by their ability to listen to different points of view, especially if they do not agree with their own (Grimaud et al. 2017).

Wu et al. (2019) discovered an intriguing phenomenon: solo authors are just as likely to produce high-impact ideas (in the top 5% of citations) as five-person teams. They are also 72% more likely to produce disruptive concepts. More generally, scientific innovation of a team grows steadily with each new member added. The organization and management of research teams are linked to this aspect, in the academic world, of a solitary inventor!

3.1.5. *The question of finalized knowledge integration*

> The notion of knowledge implies both various cognitive acts, such as observing, reasoning and communicating, and skills for carrying out certain tasks. It also brings with it the notion of 'frontier', where representational systems can come into contact, and beyond which exchanges are regulated and information flows filtered (Arpin et al. 2020).

> A scientific discipline is a contextualized, culturally oriented practice, built within linguistic, instrumental and organizational frameworks. It is always part of an architecture of knowledge production that both exceeds and integrates it (Arpin et al. 2020).

> A ruling class can rarely remain in power for long, if it is not prepared to open its ranks to new, capable and ambitious members from below (Burnham 1964)[7].

The purpose of this pooling must be seen as an attempt to integrate to try to bring out a "transcendent" point of view from the different perspectives (McMurtry 2009). Given the complexity of the processes under study, interdisciplinarity appears to be a sine qua non for better identifying and making sense of the complex (Newell 2001). "The modalities of this integration are not clear, however, and raise a number of questions concerning, for example, the prevalence of certain processes; does all the disciplinary knowledge acquired have the same weight for a concrete observation?" (Klein 2001; see also Le Fur (2014) on the same subject).

The direction of the proposals is, "without negative connotations, part of a cobbled-together tangle of reflections, a necessary scene of creation that will induce a trajectory" (Louargant 2003). "At a time when we are becoming aware of the complexity of events, phenomena and facts, isn't every discipline a matter of perspective, of a reductionist angle of view, of the incompleteness of overly categorized knowledge?" (Arpin et al. 2020). How is the experiential antecedent reflected in changes in representation prior to the acquisition of experiential knowledge? (Gambier 2019).

The approach adopted may aim to develop a facetontology-type ontology (Madin et al. 2008), in which the attempt at concretization is carried out on disjoint, or orthogonal, sets of concepts such as discipline, biological object, tool and so on. On the more scientific side, it is natural to use the panoply of tools and collectors associated with the idea. Newell (2001) characterizes the complexity associated with the idea by its interdisciplinarity, for example (but this can take time, depending on the number of variables involved). The challenge, however, is to enable articulation to reveal new processes or dynamics arising from disciplinary articulation (McMurtry 2009), for example, by bringing together different spatiotemporal scales.

Drawing a parallel between expertise and politics, we can consider, in principle a "Machiavellian" approach: dispassionate, often skeptical, but realistic. It is a method that does not judge whether a system is good or bad by simply observing it and drawing a few conclusions. Machiavellian thinking begins with the intuitively

7 Author's translation from the French version of the text.

obvious idea that all human organizations need leaders who, in Burnham's (1964) words, are "those who take the struggle for power seriously". This is indeed what we often find in meetings where the "dress effects" of some people steer the debates. In principle, these personalities possess certain specific qualities, such as broad skills (able to appreciate the pertinence of the contribution of the processes of change induced by the introduction of the idea on transformations in representations), acute intelligence and high motivation (Andreen 2021). The complexity of the idea requires a "mature" organization covering the subject, preventing all of its members from being equally involved in the decision-making or clarification process (especially if this number is high, depending on their clarification time). An analysis aimed at clarifying the idea must avoid the "iron law of oligarchy", which describes the tendency of any organization to secrete an oligarchic elite (Michels 1914).

There is therefore reason to try, in order to avoid this reality (accepting in practice the scientific domination of a minority, involuntarily or not), to play on the basis of replacing experts (see the circulation of the Pareto elite). This risk may be real, in particular because the pooling of knowledge cannot erase hierarchical relationships, the imbalance between knowledge and non-knowledge, unequal access to scientific and technical knowledge, the problem of the representativeness of the group of experts or the cognitive impasses in which, ultimately, it is a leader who sets the rules and the orientation of the expertise (Mazurek 2016).

Regardless of this aspect, for Lenoir (1995), a rationality of the choice of experts must intervene as soon as the idea can be considered as an explicitly formulated problem (while we seek precisely to make it credible). The latter must, for this author, have a good general culture. He warns against the fascination with a too loose basic training where we know everything about everything (or nothing about nothing). The training should focus on the proper use of this training, "which requires a regular reconsideration of its content in the light of scientific development and social purposes collectively retained". So, in these conditions, "it is therefore appropriate to abandon the ideology of a reality which would be the pretext of knowledge to substitute for it the utopia of a reality which would have no other consistency than that of the symbolic organization of plural discourses" (Resweber 1981).

There is therefore a debate aimed at an "optimization" between specialized skills useful for the clarification aspect and a general knowledge allowing overlaps between normally disjoint disciplinary knowledge to succeed in overcoming compartmentalization mentalities and projecting toward credible proposals. At this stage, we are not considering an undifferentiated "fusional" interdisciplinarity, but collective work (interdisciplinarity cannot be dissociated from organizational issues) making it possible to give substance to an idea (for example, to bring an apparently simple discourse, to a more complex situation that needs to be framed). It is a question here of pooling creative, logical and critical thoughts.

3.1.6. *What interdisciplinarity to call upon?*

> Nostalgia for a humanism that is both scientific and literary as the eighteenth century knew it, permanent theme of a science of sciences or of absolute knowledge, requirement of science as a result of its very development, crisis of the university, all these elements undoubtedly combine to constitute the current historical dimension of interdisciplinarity (Berger 1972).
>
> Every interpretation is based on a reserve of prior experiences, which are our own experiences and those passed on by our parents and teachers. They function as a frame of reference in the form of available knowledge (Schütz 2008).
>
> Interdisciplinarity refers to institutional and intellectual issues (interdisciplinarity is conceived as knowledge capable of grasping 'what is woven together', i.e., according to the original meaning of the term, 'the complex', and runs counter to 'knowledge that separates, breaks the complex of the world into disjoint fragments, and splits up problems'), epistemological (interdisciplinarity brutally calls into question science, at least the conception given by epistemology) (Morin 1998).

The association of partners from a variety of disciplinary backgrounds refers to open systems and the use of the project concept, leading to new epistemological attitudes: "It is no longer so much a question of grasping the object in its invariance as of situating it in its finalized evolution, through the project or projects it sets itself" (Bibard 2017). Working "inter" means sharing expertise: interdisciplinarity is a working process, not an end in itself. In this sense, knowledge disciplines can be essential to the smooth running of expertise clarification. In conventional approaches, we do not want to develop a superficial view of the phenomena under study. In this stage, it is "just" a matter of ossifying a proposal as an extension of an idea. However, in a limited space of time, mutual adjustment to each other's working methods seems necessary, with points of friction and divergence normally arising between the different approaches.

In virtually all of the previous sections of this chapter, interdisciplinarity is linked to the "tug" between the development of specialization and the need to ensure a coherent synthesis. James (2011) suggests that interdisciplinarity makes it possible to "multiply ideas and intellectual contributions while avoiding reductionist tendencies. [On the other hand], inconsistency is the main risk resulting from the expansion of intellectual diversity". Indeed, for Neumann (2011), the main obstacle to its development remains that of coherence or epistemological cohesion. For its part, in this sense, Sinacoeur (1983) considers that interdisciplinarity does not refer solely to a category of knowledge, but indeed to a category of action, hence a growing consideration:

> Interdisciplinarity betrays a characteristic of our time: the social integration of knowledge, henceforth a constitutive element of power, and power is essentially interested in applicable knowledge, the only one capable of guiding it in the formulation of the programs which articulate its exercise.

This is somewhat what is expected in the clarification exercise.

The practices of the so-called clarification operation involve encounters, dialogues, debates and oppositional tensions between positions of otherness, within an event that brings them together without confusing them. Otherness would indeed be at the heart of interdisciplinarity-process, as the driving force behind a process of discovery and exploration of the boundaries of territories to be shared (Lesage 2008; Bienaise 2016). It is naturally hoped that interactions between representatives of disciplines will produce "novel analogies or metaphors, creating new mental categories in the cognitive universes of the actors, and [...] sometimes provoking major epistemological changes, potential sources of economic growth" (Ancori 2014).

3.1.7. *Grains of sand in the clarification process*

It has been reported that a more or less interdisciplinary group has been set up to give substance to the idea, described above as "confused apprehension". For a researcher, working within this framework means working in academic structures which are not designed for this purpose. There is no special recognition to be expected, nor funding envisaged. Everything must be done more or less for free and without authorization from hierarchy because we agree to help. But, obviously, this volunteering has limits, even if the use of the Internet can be beneficial to the operation. However, the clarification requires a number of experts changing with the evolution of the refinement of the latter. But, in the compartmentalized universe of the world of academic research, how do you manage to have the necessary experts available who are expected to agree to contribute to the operation?

This major difficulty may explain why ideas leading to inventions are considered to come from isolated individuals with almost all the qualities of the group. In this case, it is a case of self-clarification! A difficulty associated with expertise stems from the need to lend credibility to aggregations of disciplines "that do not naturally or mechanically fit together, a problem that is compounded when we consider the links between things as heterogeneous as disciplines and sectors, concepts and people" (Sinacœur 1992). This may not be the case in the industrial context, however, because there are other, more powerful forces at work.

In addition, the activity, carried out, always using the same experts, risks leading, in the more or less short term, to a homogenization of the operating methods of the latter. "Good" disciplinary diversity is an initial condition that has been mentioned; it must be recognized that the constructive contributions, from one form of expertise to another, will be predictable and will eventually lose their relevance for clarification. In the context of clarification with its constraints of time pressure, the preferred mode of exchange is certainly controversy. For Ancori (2014), one reason for such habitual degradation is that interactions between experts are only constructive in a weak sense, "because the new representations they produce are never more than updates combinations of mental categories already present in a virtual state within the network considered globally".

To overcome these difficulties is to bring new blood into the (informal?) group involved in clarifying the initial idea, but with the already expressed difficulty of recruiting new members. Each addition, like each departure, mechanically affects all of the mental categories and can renew relational combinations. The second possibility, proposed by Ancori (2014), consists of introducing the production of new categories via the formulation of novel analogies or metaphors. The analogy-producing cognition thus proposed assumes the anticipated construction of an expertise, which must be free and rapid, with the sole aim of ossifying the idea.

On the other hand, if we can talk about the legitimacy of a good recruitment of experts ("Legitimacy is not, therefore, a stable attribute of certain positions of power or of certain approaches. It can only be invoked in the context of a perspective, an orientation, a project – what constitutes a project remains to be defined" (Reynaud 1997)), there will come a time when it will be necessary to evaluate the relevance of advice. This last aspect must be taken into account at the moment when the idea is being investigated, in order to go on to invention (or stop along the way). In this case, it is a matter of participating in the extension of the idea, inferring from it, and giving decision-makers food for thought. Repko et al. (2012) also raise the question of how many disciplines are actually needed to tackle the problem.

There are, however, differences in trust between people due to differences in generation, level of education, language, income, gender and region, hierarchical relationship, etc. Apart from the expertise envisaged here, progress can no doubt be made in this regard. In a framework of trust, it is possible to let ourselves be carried away by curiosity, to consider the exploration of new ideas or the lateralization toward other targets of that which is the subject of the expertise, all the more than in science, invention should be a major criterion for success. Bouveresse (2021) speaks, however, of social correctness and truncated opinions. He writes:

> We must therefore ask ourselves if the real deception, the one that is in a certain way at the origin of all the others, is not precisely critical deception, in other words the behavior of a criticism – view most of the time of all kinds of sense, distance, and even more critical will. And who finds it both normal and natural to do almost the opposite of what is expected of her most of the time.

Not only does this unethical carelessness not only fails to support the idea, but can also lead to significant financial damage. And then, if the idea goes all the way to innovation, you can always say to others: "I was there!".

"Disciplinary diversity expresses itself as much as a strength as an obstacle for work teams, as it can also lead the group to conflictual dynamics, delays in deadlines, stagnation of actions, a climate of mistrust and results inappropriate to objectives" (Chauvey and Georgescu 2017). However, there may be cognitive biases which are errors in reasoning. When we speak of "error" here, we do not mean that cognitive biases deviate from the standard model of rationality. Can we then systematically rely on our estimates and snap decisions when data and time are limited? What is more, every decision-making situation generates uncertainties and tensions. In this context, complexity is often unavoidable. Claessens (2016) believes that this reality is likely to be beneficial in terms of the proliferation of initiatives it allows and the autonomy of the players it can foster. "How, then, can we distinguish 'good' complexities from 'bad' complexities?" (see in response Poutanen et al. (2016)). For Giget (2020), we need to add the need for "weapons of mass reduction" to enable us to move forward without betraying the idea too much!

Systemic modeling is not an easy operation for researchers to integrate. In direct confrontation with a complex situation, the researcher finds themself faced with several difficulties of perception: partial, even biased analysis of the situation to be studied with fuzzy boundaries of the system of interest, unstable nature of the system of study, taken apart from its environment which leads it to adapt, etc. But, as noted, these areas may not be covered due to lack of time. The multi-dimensional character of the real problems hidden behind the idea can challenge practices such as those of systems engineering. At the stage of clarification, it is necessary to overcome the difficulty of defining the problem to be solved without restricting its globality and complexity, in particular in the first stages of the life cycle (Lizarralde et al. 2011).

A clarification is said to be robust if the response is little modified by perturbations (time of expertise, number of experts, their origin, etc.). Insofar as it is possible to do so, different clarifications with different conditions may be very close (robustness) or not. The result is a certain fragility of the opinions expressed, if only because it is sometimes difficult to endorse arbitrations. However, if we take the example of Switzerland, ranked world champion for innovation, the academic world does not seem to reflect what is happening in the economy with modest support for applicable research, called "service to the city" (what about France?):

– risk aversion;

– low commitment for learning;

– low openness to new ideas;

– cultural traits that maintain horizontal rigidity;

– cultural traits that maintain vertical rigidity;

– low valuation of success and negative sanctioning of failure;

– performance orientation with a short-term vision (Boukamel and Emery 2018).

The organization of research must consider the complexity approach as one of its areas of focus. It can be destabilized when research requires the flexible collaboration of several disciplines. In such cases, the production of scientific knowledge needs to move away from its usual organization to address practical problems. This situation goes beyond interdisciplinary scientific confrontations.

The research community sometimes takes a critical view of scientific policies and regularly refers to the lack of clear positions taken by the authorities (Classens 2016). Otherwise, we run the risk of finding ourselves in the situation of Rabelais' Abbey of Thélème, whose rule was: "Fay ce que vouldras", provided that everyone does the same" (Weil and Dubey 2020). On the industrial side, Galano's (2020) opinion is hardly more enthusiastic due to the intermediation of a "transverse" manager who is appointed to manage a project, for the reasons outlined as follows:

– The cross-functional manager has an obligation to deliver results.

– The only way to get them is to mobilize cross-functional human resources for the benefit of your business, but from other departments.

– These resources must perform other activities, whether they are also transversed or serve their own function.

– The line manager also has results targets to meet.

– They are also judged on the efficiency of the human resources in their charge.

– They also have the formal legitimacy to set objectives for their employees and give them the means to achieve them, primarily their own time.

– They also have the formal right to appraise their staff in terms of their achievement of the objectives they have set for them.

– The cross-functional manager has no formal authority to evaluate contributors.

– Faced with their statutory inability to obtain the contributions needed to achieve his results, the cross-functional manager initially uses means of mobilization and persuasion.

– It can achieve its aims, most often when the project or activity is designated as a priority by senior management.

– But not always: "temptation or despair can then lead them to implement coercive mobilization strategies, notably by creating insecurity, by displaying hostile or authoritarian attitudes, thus transforming the conflict of objectives into a conflict of persons" (Galano 2020).

On the other hand, because of its novelty, its management aims at "fuzzy" aspects, because the cooperation processes that take place are weakly structured, and the confrontation of the different actors' points of view is essentially supported by informal exchanges (Lizarralde and Pilnière 2012). Arrow's "impossibility theorem" (1951) also teaches us that two actors, faced with a number of possible choices equal to or greater than three, cannot find an exact match for their preferences. This situation leads to the imposition of constraints on the experts, which can lead to tension within the group.

3.1.8. *Conclusion*

> The challenges of a disciplinary approach are often correlated with the complexity of objects of study, which eludes the understanding of a single discipline in the production of knowledge. Those linked to the inter-disciplinary approach are, first and foremost, the divergence in the process of integrating disciplines that are relatively 'relevant' to the objects of study (Repko 2012)[8].
>
> A new paradox indeed strikes our scholarly institutions massively: if interdisciplinarity is today an omnipresent watchword, it is at the same time repressed and feared. Everywhere we swear only by collaboration between disciplines, and everywhere specialists refuse to open up to this collaboration (Barthélémy 2013).
>
> People, when they think, have an unfortunate tendency not to hear what bothers them (Revel 1988).

Creative or innovative personalities are apparently hungry for opportunity. In the company, they can have a favorable environment (this is not systematic). In the

8 Author's translation from the French version of the text.

academic world, creatives are not specially supported by their organization. Basically, the academic researcher is alone or lives in small teams (even if the research unit is large, with a few hundred people). They cannot easily benefit from the support of a light expertise for a clarification of their idea. After all, it is counterproductive for these opportunistic teams to be overwhelmed by opening possibilities without being able to participate in them (especially when it takes several years for a patent to be taken): this is one route to anti-invention, the other possibly coming from the evaluation of the researcher (Muller 1993).

The clarification operations involving interdisciplinary approaches correspond to an "instrumental" approach whose aim is to solve a concrete problem of ossifying an idea, and where "interdisciplinarity is perceived as an essentially political practice, that is, as a negotiation between different points of view, in order to finally decide on a representation considered adequate for action" (Fourez 2001). Goal orientation can lead to inattentional blindness, "i.e. the inability to perceive things that are perfectly visible [...]. Patterns limit attention, determine the limits of actuality, and they direct attention to pre-established targets" (Bonnardel and Lubart 2019). This situation can be reinforced because complexity implies translations between different languages, which adds an extra degree of uncertainty to the use of this notion.

The representation can influence the work of the experts involved in a potential clarification. As the analysis progresses, if the complexity of the of the perceived subject increases, it becomes difficult to produce effective representations. For Casakin (2007), in a complex clarification, there may be doubt about the correct use of applicable or necessary concepts, rules and principles. Complex situations are organized in such a way that they present a variety of solutions, many possible courses of action or no acceptable solution at all (see Figure 3.6). However, experts need to be able to distinguish between the various possible solutions, so that they can quickly eliminate nonsensical proposals. "We can therefore content ourselves here with summarily describing the content of stupidity as any manifestation of attachment to derisory themes, which are inexhaustible both in number and variety" (Rosset 1984).

Interdisciplinary projects, involving independent experts, must accept honest confrontation with other specialists and the affirmation of the heuristic value of controversies (without "dress effects"). It is therefore desirable to have, whenever possible, the most accurate and reliable of the proposals resulting from the approach taken by the disciplines involved in the project. Interdisciplinary expertise is, as already mentioned, a delicate process involving the right choice of disciplines, experts and organization to achieve the clarification objectives with shared learning. This activity can be programmed by companies, but probably not in the academic world. Indeed, Sevtsuk et al. (2022) show that campus planning and building design, to intentionally create daily walkways that bring researchers closer together seem, along with cafeterias, to be effective for MIT.

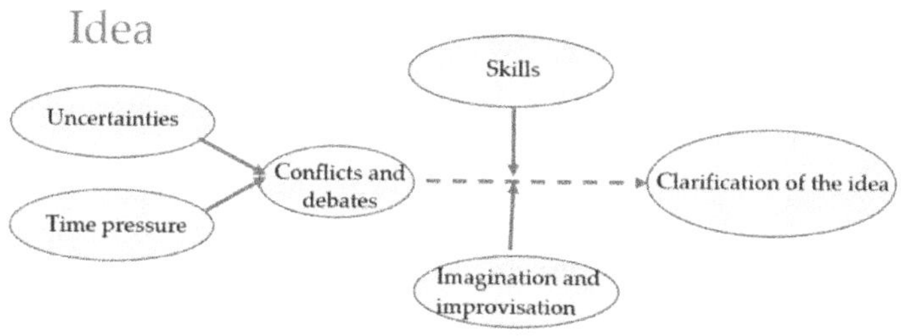

Figure 3.6. *From idea to clarification*

Some researchers, speaking about the workings of academic research, point to the constraints and sometimes obstacles that limit their work: difficulty in obtaining funding, evaluation, career, positioning within the research structure, etc. This discomfort at work can lead to marginalization or submission to the research management system. Veissier (2022) has shown the weight of the doctorate in French research. To some extent, academic researchers carry out their research "by proxy". Insufficient supervision, lack of respect for the commitment to research training can also limit the emergence of radical ideas. Fortunately, this is not always the case! In these extreme situations, help with clarification (which we do not observe in companies motivated to develop their inventiveness).

In general, disruptive innovations are faced with several difficulties related to the particular nature of their object:

> The first is that a disruptive [invention] cannot be decreed; the second is that a disruptive [invention/innovation] takes a very long time to have a significant impact in terms of turnover and the third is that there is a high risk that the entity will quickly become isolated from the rest of the organization (Silberzahn and Rousset 2019).

But, following the clarification operation, the selection of disruptive ideas is neither easy nor intuitive. It is difficult to support a possible development if only because, in companies, the invention may require costly modifications (obsolescence of processes). There is therefore a risk of seeing radical innovation ideas rejected (Bessant et al. 2010). Without having the views of partners (if they can be found), there is a risk that promising ideas will be rejected due to time constraints and, without traceability, they will disappear. At universities, with the functioning of supervised calls for funding, there is a tendency to "flutter", following the calls for proposals, which constitutes an opportunity for creativity, but we do not have data on this subject. In general, it is necessary to have strong values to change traditions through disruption. However, the path is not marked out...

3.2. A comparison with 3D, 4D and bio-printing

> If everyone thinks the same way, someone isn't thinking (Patton 2017).

> Leadership is the art of getting someone to do something because they want to do it (Eisenhower 2018).

> At the moment of work, when little by little an understanding begins, takes shape, deepens; when in a confusion little by little an order appears, or when what seemed familiar suddenly takes on unusual, then disturbing aspects, until finally a contradiction bursts forth and upsets a vision of things that seemed unchanging – in such work, there is no trace of ambition or vanity. What then leads the way is something that comes from much farther away than the 'self' and its hunger to constantly expand (be it for 'knowledge' or 'knowing') – from much farther away surely than our person or even our species (Mumford et al. 2017)[9].

As illustrated in Figure 3.7, disruptive innovations are generally characterized by activated or driven technology, limited initial functionality (lower performance), new or different functionality (different attributes/point of difference), a different price/cost structure (lower margin) and the introduction of new customers to the emerging market (Montgomery et al. 2018). It is easy to imagine how attractive an invention can be when it offers performance that traditional technologies cannot match.

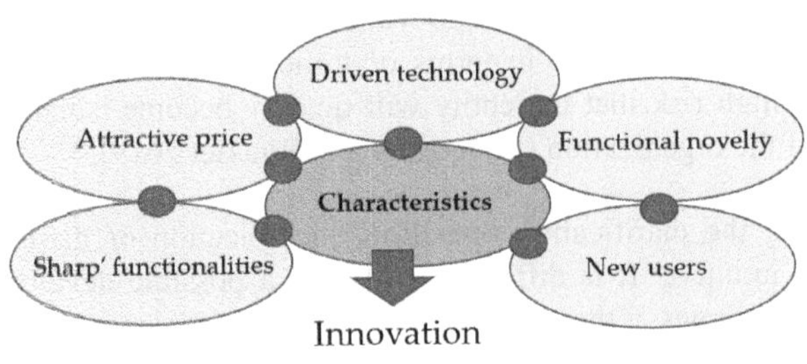

Figure 3.7. *Properties of disruptive innovations (source: Montgomery et al. 2018)*

The three technologies that will be mentioned have the same source, the invention of additive manufacturing in 1984. While the first corresponds to a

9 Author's translation from the French version of the text.

process that is simple to implement in principle (just in principle), the other two envisage refinements: this is 4D printing, in which so-called active materials are incorporated into the object created by additive manufacturing. Under energetic stress, these materials enable changes in form and/or functionality, while bio-printing is the additive manufacturing of living matter for personalized medicine. 3D printing is now an industrialized process. Below, we examine whether or not the sophistication found in the other two technologies can be used for applicative developments.

3.2.1. *Additive manufacturing*

> It's because his head is empty that he deploys this wild inventiveness and it can operate where others 'can't' work, because they're caught up in the flow of incoming information that they have to process and render at full speed (De Casabianca 2019).

> We have entered the 'era of *Homo Creativus* as an economic player'. The individual is expected to think creatively, mobilizing cognitive skills specific to creativity such as 'flexibility' to be able to 'apprehend a single object, a single idea, from different angles' and to 'disengage from an initial idea to explore new avenues' (Lubart 2010).

> The concept of confidence has a strong content: halfway between the faith of the coalman and systematic skepticism, secure enough to be used as a basis for practical action, trust is an intermediate state between knowing and not knowing about others. This confidence takes two forms: calculated confidence and affective confidence (Arasse 2000).

3D printing or additive manufacturing is presented as a central element of a new industrial revolution in which digitization, information and connectivity are transforming innovation in the creation of objects. However, if the supposed advantages of this technology are convincing (annual growth of the order of 20%/year), existing research is of an incremental nature, focusing on digital aspects, processes and materials. But, initially, Alain Le Méhauté's solicitation was quickly translated into a simple solution (the old "KIS" precept for Keep It Simple), recounted in André (2017a). In this operation, experience knowledge was, of necessity, the unavoidable trigger for 3D action. So, historically, invention, and then innovation, are part of what Avenier (2019) calls a "design without an end goal" in which the ends have not given rise to changes in the means.

Spatial resolution for light and light-curing resins were all it took for the first 3D technology to be born. The principle additivity, as shown in Figure 3.8, was designed. However, while this was not considered in the first tests, each elementary volume (voxel) adheres to those already manufactured (the idea of printability). Since then, other 3D technologies have been developed, enabling the creation of objects from polymers and biological materials to ceramics and metals. The principle of additive manufacturing is generic and, apart from material savings, offers many advantages over subtractive manufacturing processes.

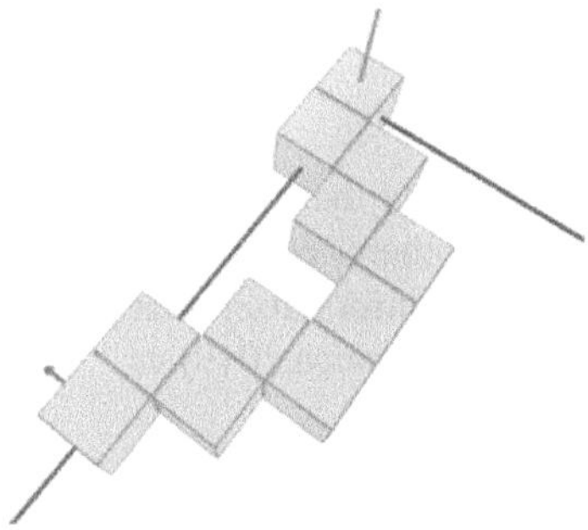

Figure 3.8. *Additivity principle (a reminder)*

On the one hand, the proof-of-concept could be carried out by a single person, without the need to set up a group for possible clarification over a weekend and, on the other hand, the interest of this new process was immediately understood by the industrial world. The "magical" success of the operation was followed by a reflection on the possible follow-up to this little work. For the author, amplification also came from the media (television, creativity awards, etc.), which were effective relays for promoting the idea. The invention was exploited rapidly because it was accepted, given its potential performance (see Figure 3.9), which was superior to that of subtractive methods. However, as is often the case, real economic development took around 20 years to assimilate. Today, the market is worth around 30€ billion/year, growing at a rate of 20%/year (with subtractive technologies, which have proven their worth to a large extent, still in use).

In this operation, the transfer of research technology from the public sector to industry can be analyzed on the basis of the so-called "linear" model of innovation, reflected notably in Bush's (1945) report, which assumes that public sector research is applied in industry in a unidirectional way, without active participation.

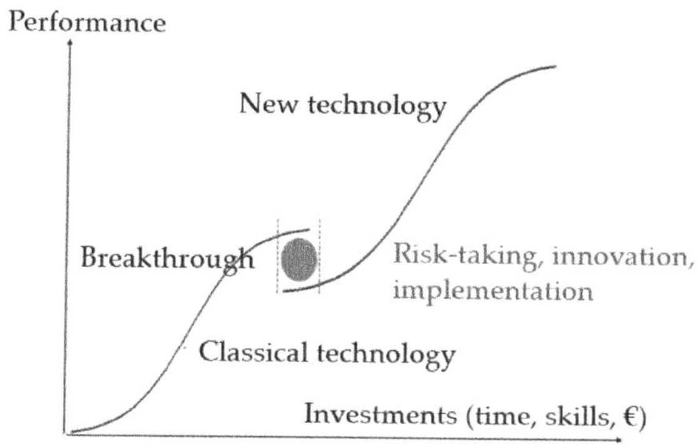

Figure 3.9. *Attractiveness of additive manufacturing*

As Horenstein (2002) explains: "If there is only one answer to the problem, and it's simply a matter of putting the pieces together, then the activity is probably analysis[10]." As mentioned above, this is what the "DIY-maker" does, bringing together scientific and technical elements to integrate an idea of the whole to correlate numerous ideas and information from various fields. It is a direct disruption that leads to invention, then, through attractiveness to innovation. Since this adventure began in 1984, research has been carried out into the relationship between materials and processes to obtain industrial parts with the desired properties (hence the progress mentioned). This field of interaction is part of the demand from users of 3D machines. It is partly linked to the existence of 3D manufacturing laboratories with several types of machine offering a wide variety of possible applications.

From a bibliographic point of view, the diversification of materials continues. This comes from the possibilities of developing new materials from usual constituents, but also from the need to optimize the properties of use. Thus, beyond composites with a metal or polymer matrix, we are seeing the emergence of green composites, nanocomposites, "intelligent" materials (biomimetic materials, alloys/shape-memory polymers, piezoelectric materials, adaptive micro/meso-structures, etc.), architectural materials (nanometric/millimetric, cellular reinforcements, foams, porous, granular, fibrous media, etc.), metamaterials, without forgetting biomaterials or living tissues. These "new" materials, most often "carriers", give rise to new work, in particular on the "printability" aspects.

10 Author's translation from the French version of the text.

NOTE.– Choosing the "right" material a priori.

Based on knowledge of 3D machine operating modes, commonly used materials and results, several knowledge-based methodologies (weaknesses in rationality) can be proposed to define a "correct" choice of the right material(s). To do this, we use our own (so-called long-term) memory, which stores what we know from personal experience, with the rest coming from literature and contacts with people working on similar themes (including Fab-Labs). For beginners, it is easy to imagine how difficult it can be to obtain the best possible "materials" information, especially if you are aiming for more sophisticated 3D productions.

Through bibliographical knowledge, it is possible to obtain information that is more or less stable, and a publication-by-publication analysis is not only tedious, but requires a summary that is difficult to carry out (especially for the uninitiated). The latter can be found in books or review articles (with the risk of a certain obsolescence of the data). In such cases, the names of the materials and some of their properties may be available, but the "practical" aspects are rarely presented in this literature (at least for lack of space). So, in an initiation process, it can be advantageous to exploit the most restrictive criteria to avoid a combinatorial explosion of possibilities, thus keeping the design task within reasonable time limits. However, this raises the question of the knowledge of a methodological approach that would take into account criteria of printability costs (material, process, etc.), object properties (e.g. Young's modulus), material recycling, environmental considerations, etc., with a (proper) hierarchy of these criteria.

It can then be useful to play with highly selective criteria to avoid impossible choices or an excess of possibilities. If the discipline of constraints is too rigid, many possible development directions will be closed, which can be advantageous (single solution), but with the risk of low flexibility. The latter allows new knowledge to be used as soon as it emerges (which is not what the "future (?) practitioner" is looking for at this stage). It can respond to new criteria whenever they are supported by scientific and technical literature (alas, not everything is accessible). It can also be used to broaden and/or better specify objectives and introduce new selection criteria.

NOTE.– Owning an additive manufacturing machine imposes most selective criteria for materials. Machine manufacturers provide lists of materials, which saves a considerable amount of time. However, this does not allow you to go much further.

Additive manufacturing continually offers new production methods that incorporate greater design freedom. It relies on adaptive, agile and cost-effective methods, enabling faster and cheaper production of good-quality, customized parts under mass-production conditions, with gains in mechanical performance and resources used (Hahn et al. 2014; André 2020). The industrial applications of

3D printing concern the value chain in many fields: prototype parts, energy, healthcare, mechanics, chemical reactors, exchangers, industrial equipment and tooling, aerospace, avionics, automotive, transport, construction and public works, food, etc. The size of objects from a few nanometers to several tens of meters.

3.2.2. *4D printing*

> The inevitable never happens, while the unexpected always does... (Keynes 2016).
>
> Let's not mince words. This 'engineering' is a modest DIY that is much more akin to the empirical processes of the amateur gardener than to the reasoned knowledge of the agronomist (Sérieyx 2018).
>
> There is no simple idea, because a simple idea [...] must be inserted, in order to be understood, into a complex system of thoughts and experiences (Bachelard 2003).

Based on advances in additive manufacturing, 4D printing emerged in 2013 to give apparent life to 3D objects, exploiting additive manufacturing techniques and so-called active or intelligent materials. From a bibliographic point of view, we find, as shown in Figure 3.10, a background corresponding to this exciting and new field. It is interesting to examine how to use knowledge of "morphogenesis" (Turing 1952) and other ideas of self-organization to achieve the desired form to exploit different physicochemical systems whose (at least spatial) structure can lead to changes in the form or functionality of objects. However, the solution to the inverse problem, applicable to direct manufacturing, has yet to be found.

In this field, where complexity reigns supreme, we classically return to the pillars from which we know how to go further: classical additive manufacturing with heterogeneous and/or transformable materials. The range of metallic and/or molecular systems, with no clear preference for any particular approach. In fact, there are references to some proofs of concept in an emerging field that probably overshadow a number of constraints technological constraints (Demoly and André 2021a, 2021b, 2022a, 2022b).

By modifying the environment, in the presence of electromagnetic, photonic, chemical, biochemical, mechanical fields, etc., it is in principle possible to create and then change the shape of objects and even their functionality(-ies). This is the basic idea behind 4D printing. Work at the Massachussetts Institute of Technology (MIT) (Tibbits and Thomas 2013; Tibbits 2013, 2016; Thomas et al. 2014) initially made these basic concepts a reality. This idea, highly original with its demonstrator, serves as the starting point for an emerging scientific field (44% annual growth in the number of scientific publications) (Demoly and André 2022b).

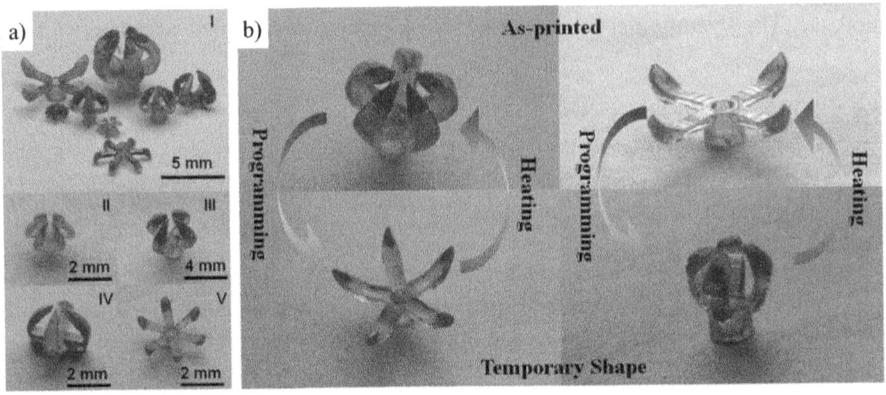

Figure 3.10. *The anisotropy of the material creates a curvature when changing temperature. Complex morphologies (here a flower) are generated with evolutions in object shapes (source: Ge et al. 2016)*

However, to the author's knowledge, no clarification has been carried out, despite the fact that, unlike 3D printing, whose founding concept is very simple, the number of parameters that can be played with is particularly high: 3D processes, materials, multi-materials with their spatial resolution, mode of stimulation, spatiotemporal distribution of stimulations and their amplitudes, etc. So, as in Bruegel's hay wagon, each of the scientists involved in exploring the theme carries out a localized piece of work, without considering the generality of the idea.

On the other hand, while the proofs of concept proposed by Tibbits and its successors are attractive (adding magic to magic!), it is possible to show that with the results in the literature, it is not feasible to produce demonstrators with industrial potential: low Young's modulus, high response time, uncontrolled localization of active zones and their stimulation, etc. (Tan et al. 2018, 2019; Demoly and André 2022a, 2022b). We then find ourselves in the situation shown in Figure 3.9 (difficulty of transferring the 4D printing idea), where we have an idea that represents a break with what already exists, without its applicative realization having acceptable performance relative to what already exists.

Subject	Comments	Ranking
Adaptive optics	Small-amplitude mirror displacement	
Space engines	4D metal deflector	
Work	Adaptive seats	
Comfort	Self-adapting pillow/nose plug	
Everyday/education/play	Book whose pages change as the story unfolds; learning to write; letters appear and disappear to guide children through the story	
Baby game	Example: a cat in water turns into a duck	
Game: 4D lego	Evolving structure	
Hygiene	Bathtub: water volume versus fill level and body shape	
Health & Safety	Easy-care 4D bactericidal sponge: becomes smooth for easy cleaning	
Chemical hazards	PPE that closes/opens in the presence of a chemical product	
Baby	Fever-detecting pacifier	
Medicine	Regenerative medicine	
Medicine	Adaptive implants	
Flat storage	Various equipment (kitchen, construction, etc.)	
Sex robot	4D actuators	
Smart robotics; exoskeletons	4D actuators	
Adaptive seat	4D actuators	
Disability	Staircase that transforms for accessibility	
Disability	Garment that transforms in the event of loss of balance to prevent falls or to lift a fallen person	

Subject	Comments	Ranking
Home automation	Floor becomes nonslip when wet	
Home automation	Table surface that sinks when liquid is spilled	
Anti-theft devices	4D marking to complement other	
Signage	Signage that changes according to the environment and context (rain, day, night, noise pollution, etc.)	
Adaptive gloves	Texture changes with effort	
Adaptive footwear	For example: crampons appear if there is snow	
Safety sport	Unobstructed protection of athletes' movements (skating, gymnastics, rollerblading, etc.)	
Airbags	Adapted "passive" 4D shapes	
Security	Garment that turns into a buoy on contact with water	
Security	Auxetic materials that break when hit by a projectile	
Personal safety	Adaptive floor that alerts when you fall (weight effect)	
Works of art/jewelry	4D distortion; color change	
Decoration	Sensory and aesthetic experiences/4D art	
4D lego	4D distortion	
3D object with scalable coloring	Photochromic, electrochromic or thermochromic materials on an object	
Acoustics	Flexible fractal structure (from a flat structure)	
Projectile stop	Auxetic structures	
Drugs	Delivery localized in time or space	
Medicines	Pill common to several drugs with staggered dosing times	
Odontology	Tooth repositioning help	

Subject	Comments	Ranking
Drones (obstacles)	4D structure for folding in the presence of obstacles	
Trochoidal systems	Convertible propeller blades	
Adaptive prosthesis/orthosis	Prosthesis for children evolving with the child's body – adaptive Boston(s)	
Optics (glasses)	Lenses that evolve with user needs	
Transport	Foldable/reconfigurable/reusable vehicle – convertible car	
Watering	Automatic system	
Spatial	Miscellaneous unfolding (mirrors, antennas, etc.)	
Energy collectors	Piezoelectric	
Clepsydre 4D	Potential energy recovery	
4D stents	Adaptive materials	
Orthopedics/sports	Adaptive materials and 3D shapes	
Chemical reactors	Adaptive walls and valves	
Boat sails	Spoilers in adaptive aerodynamics; speed-dependent deformations	
Tires	Wheel torque – Tread	
Mechanical – Airbags	Damping/metamaterials – 4D airbags	
Auxetic facades	Reconfigurable shapes adapted to heat and/or light	
Clamps and actuators	Deformation of a 4D actuator object by external stimulation	
Color change	Transformation induced by electromagnetic field, light or temperature (automotive, home automation, art) – 2D active network	

Subject	Comments	Ranking
4D and structural electronics	Integrating electronics into objects (4D) including actuators; flexible printed electronics	
3D food	Food and food processing	
Liquids	Container that changes volume when there is less liquid (water, wine) to take up less space or for storage	
Anti-alcohol glass	Water/wine coupling in an adaptive glass	
Packaging	Smart packaging	
Flexible electronics	Integration; flexible printed electronics	
Electronics	Self-learning/self-repairable chips Adaptive connectors	
Processes	Heat exchangers; filtration; valves	
Industry	Adaptive clutches	
Nuclear	Adaptive safety	
Shoe	Laces tie themselves	
Objects	Self-cleaning object	
Electronics	Hydrocarbon molecule that changes form under the action of an electrical potential (insulator to conductor)	
Thermosensitive skin in robotics	Thermochrome exposed to temperatures above 50°C	
Flexible robotics	Biodegradable	
Vehicles	Foldable or reconfigurable (available from manufacturers)	

Table 3.1. *Possible application niches for the short term (green), medium term (orange) or long term (red). For a color version of this table, see www.iste.co.uk/andre/knowledge2.zip*

In this field, where complexity reigns, designing a product requires architecting not only the sub-assemblies, but also their combinations, interdependencies and interfaces, so that it best fulfills the function for which it was imagined. In simple terms, the priority here is for designers to develop a functional product, and then for methods engineers to adapt the manufacturing processes to the product. It is clear that there is a lot of work to be done. But let us hope that this new field will try to go even further, particularly by working on new materials, new structures, etc. However, in this disordered space, it is worth examining whether profitable niche applications can be explored. Based on the results of the literature (Table 3.1), which draws on that which was published in Demoly and André (2022b), and on brainstorming meetings, highlights the existence of niches (see previous reference for bibliography) which could, after instruction, lead to industrial applications.

Stengers and Prigogine (1979) have highlighted the richness of the time parameter in systems far from equilibrium. It brings a contingency to the effects of a state that no prior law can predict, and this is probably applicable to 4D printing. Each state of the object is rich in a certain past, so its future necessarily depends on it, without being able to define a correlation. We need to find the influencing parameters, even though our knowledge of them is necessarily finite. In the author's view, this situation of "rescuing" a beautiful idea, undoubtedly too complex in its implementation, should be pursued. In fact, generic activities are still to be carried out, as they could be used to develop 4D activities of interest to society. 3D printing was born in 1984, Tibbits' publications in 2013! That is a gap of 29 years, and this time, building on 3D achievements, could contribute to the desired progress.

In business, managing disruptions requires a balance to be struck, without too much of a risk, between exploiting a traditional business that is profitable, at least in the short term, and exploring areas that are uncertain but could have long-term benefits (O'Reilly and Binns 2019). This is not yet the case for 4D printing, even if promising niches have been proposed (see Table 3.1). What could happen to launch the technology is for a bold competitor to take the plunge. The robustness of the technology in its niche must then, imperatively, convert the 4D ideas concept into meaningful, readable and profitable activities. It is a job in itself, but perhaps we should consider the reflections by Edgar Morin:

> Let's say right now that a complex science will never be worth its weight in gold for the manipulative power it provides- on the contrary. [...] By enriching and changing the meaning of the word knowledge, complexity calls us to enrich and change the meaning of the word action, which in science as in politics, and tragically, when it wants to

> be 'liberation', always ultimately becomes 'manipulation' and 'enslavement' (Morin 2006).

3.2.3. *Bio-printing*

> Excited by his project, his first practical step was retrospectively: he had to turn to an already existing set of tools and materials, to make or remake them [...]. All these varied objects constitute his treasure trove, he questions them to understand what each of them might 'signify', thus helping to define a whole to be made, but which will ultimately differ from the instrumental whole only in the internal arrangement of the parts (Lévi-Strauss 1990).

> Man only escapes from the laws of this world in lightning flashes. Instants where everything stands still, instants of contemplation, of pure intuition [...] It is through such instants that he is capable of the supernatural (Weil 1991).

> To accomplish anything worthwhile in research, you necessarily have to go against the opinions of others (Livio 2017).

Bio-printing is a recent additive manufacturing technique (a dozen years old) that enables three-dimensional printing of successive layers of cells on bio-matrices in order to identically reconstitute the structure of an entire organ (André 2017c). To create life of partially artificial origin that simulates what nature can do, and even go beyond it, we would in principle need to knowledge of "natural" life. However, it is possible to imagine that, starting with "rudimentary" concepts, we will gradually rise from laboratory tests to repairs of biological tissues (Parrochia 1994).

Bio-printing belongs to life engineering or bioengineering: it aims to manufacture or repair living organs. It constitutes a kind of missing link between the artificial and the natural (see Figure 3.12). This definition calls for a comment: bio-printing cannot be adapted to all living systems, but to personalized and specific situations (of a tissue, for example) exploiting artificial systems. It is not therefore there is no question of returning to this writing by Descartes (1950): "the rules of mechanics belong to physics, so that all things which are artificial are therefore natural".

By making such constructions from our own cells, avoiding rejection, bio-printing envisages a gigantic scientific and medical adventure, with a potential market of more than several thousand billion €/year, if only for organ replacement. The basic idea and the market are there. Some scientists are quietly talking about a new, fascinating and slightly worrying world: that of stem cell engineering, biomaterials engineering, biomaterials, micro-patterning and bio-printing, bioreactors and life modeling. They assert that we are already partly capable of rebuilding our organs, liver (Samsara 2016), kidneys, skin, heart and evenbrain from our stem cells and in 3D printing, and explain how scientists all over the world are making giant strides in this field of bio-engineering, bio-printing and the reconstruction of the living... .

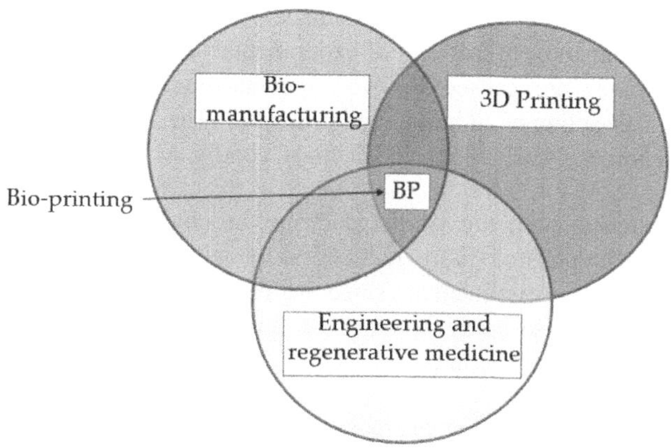

Figure 3.11. *Positioning of bio-printing at the interface with other disciplines*

As with 4D printing, why not evoke the spatial transformation of a bio-construct containing living cells energy, whatever its form or nature? We could start with a given shape, easy to produce using 3D printing, and transform it, through the addition of nutrients and an appropriate composition (biochemical energy, for example), into a shape corresponding to an application. By playing on the spatial distribution and amplitude of this energy (or these energies), it would also be possible to have an adaptive system like a "piece" of living organ. No doubt we are trying to apply forms of determinism which would assume that everything is given with the knowledge of initial conditions alone.

On this basis, the engineer's initial vision was to keep things simple, if possible within the framework of causal thinking (Hempel 1966), with the tools at their disposal, and to reject revolutionary concepts that were too far removed from their own culture. Living matter can be ordered, meaning that its form can be described geometrically, at least to a first approximation. Structures, more or less self-organized, may be associated with certain types of physical or physicochemical conditions. From a broader evolutionary perspective, it is impossible to predict the evolution of forms by deducing them from the application of simple laws that we may otherwise have been able to apply. More to the point, experience has shown that while these laws are necessary, they are not sufficient – complexity, be damned! They form the incontrovertible background to an evolution that develops in a non-apparently deterministic way, through the interaction of natural systems with each other and with their environment. The sciences of "morphogenesis" would therefore only be valid within the limits of macroscopic physics. But the latter is only an approximation, forcing us to chase linear determinism.

As far as this field is concerned, Bognon-Küss (2012) cites Claude Bernard (1865), who excluded vitalism from the field of experimental science because he asserted that vital manifestations are caused by "the spontaneous, effective and, as it were, voluntary and free action of an immaterial principle". The vital force can be associated with other phenomena, such as physicochemical conditions and so on. She writes:

> Consequence: with the vitalist doctrine, as characterized by Claude Bernard, the physiologist is reduced to the status of a mere spectator, he is not an experimenter (since he cannot act on the causes of phenomena, these being immaterial), and the physiological sciences are conjectural, not certain.

Thus, for Hempel (1966), vitalism is not only the thesis of the irreducibility of biology, a kind of "heuristic maximum", it is also the thesis according to which there are "factors belonging to another order than the physicochemical factors in the living world".

With or without metaphysical questions, we have to invest the subject with machines and processes. The literature cited in André (2017c, 2021) refers to our knowledge of the current applicative potential of bioprinting technology, with a number of promising niches (situation close to 4D printing). But, relative to conventional additive manufacturing techniques, the printing of biological elements adds a very high level of complexity to the processes because it is necessary to "intelligently" structure living or nonliving materials, minimizing the extracellular matrix and to control the spatial distributions of different types of cells or biomolecules that can play a role in cell differentiation, growth or apoptosis, etc.

To make the effect of complexity and bifurcations visible, and to better describe the overall development of a biological system, Danzo and André (2020) have used a two-dimensional connectionist "cellular automaton" model, characterized by three basic components: a network, an activation rule and a learning rule. The network is made up of a set of cells initially placed at a given point in space. Figure 3.13 shows the results of two simulations of populating cells A and B (A can be transformed into B), initiated on identical bases, but leading to very different results. It is therefore possible that, spontaneously, the causalist assumptions cannot be applied.

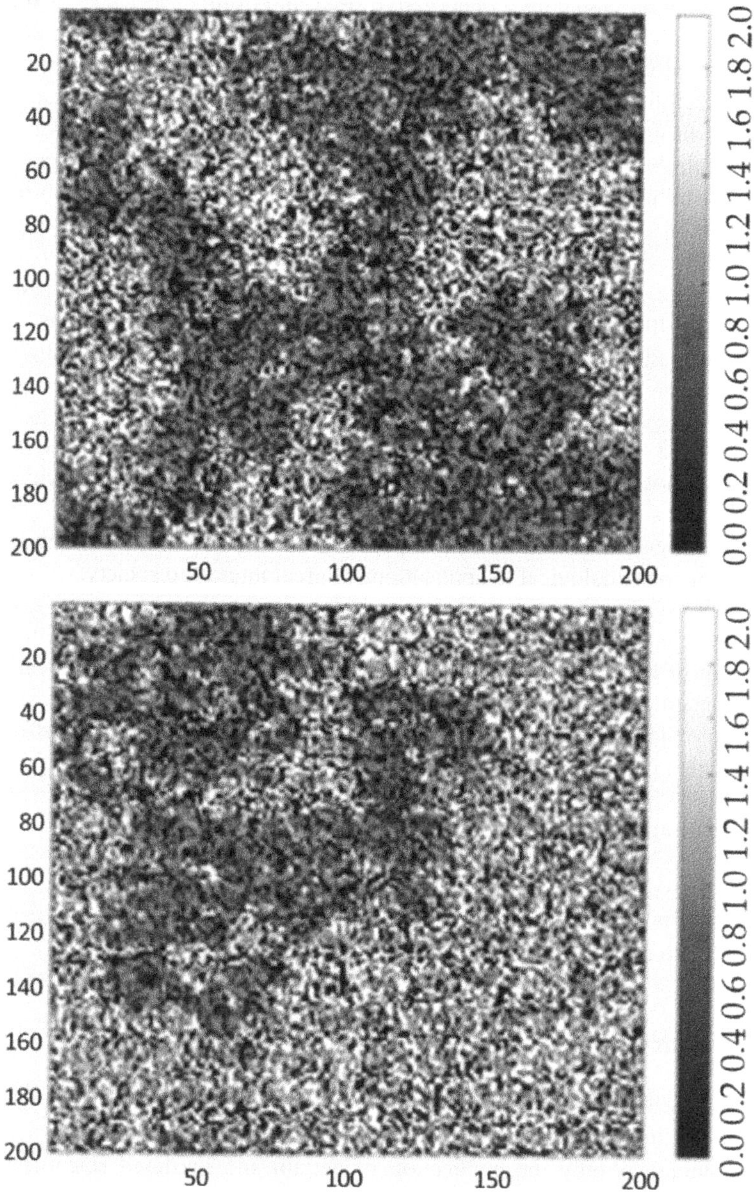

Figure 3.12. *Simulations based on the same dataset leading to different results*

Presented as an eminently collaborative field, bio-printing, because of the need to integrate principles of convergence (MIT 2016), offers a particularly interesting arena for interdisciplinarity. On this basis, after learning a shared communication language, disciplinary alliances should make it possible to reflect on the best ways to explore this promising field, although we cannot know today whether we will be able, in certain biological fields, to define a causal relationship between project (of repair, and why not of human augmentation) and expected functionality. In fact, the consultation of different partners from different disciplines must be expressed through a diversity of currents revealing the dimension, the complexity and the tensions of the emerging field that is bio-printing. This normally results in "cooperation" difficulties between highly heterogeneous players, but this may be overcome, provided the will is there, allowing us to examine whether the difficulties lie elsewhere (or not).

As with 4D printing, the space between the founding idea and its possible realization is not covered, with proposals for niches. However, exploiting a field where engineers and biologists (plus other disciplines) have to "play" together represents a rare scientific, technological and medical opportunity, but with impacts ranging from epistemological to applications of great interest to society.

> Any teleology must be excluded from the study of living bodies, in favor of the deduction of organic functions from the sole mechanical arrangement of the parts, at least any teleology which implies the intervention of souls or of some spiritual principles that that is, as true efficient causes of the movement of the parts (Descartes 1996).

NOTE.– Niche applications.

With non-vascularized tissues produced by bio-printing, Hanaphy (2022) reports the success of a corneal graft on a rabbit (before considering it for humans). Niches exist in order to make progress.

3.3. Conclusion

> Engineering and, in general, the work of engineers, seem to have to evolve between two extremes: that of rigor and certainty, where technique is only the application of certain and mastered scientific theories; and that of compromise and plausibility, where the technique is an empirical and concrete knowledge approaching a complex, plural and diverse reality, which no theoretical and scientific rationality can globally embrace in its entirety (Bachimont 2005).

> One of the hallmarks of creative places is the right to go beyond hierarchies and to show a certain irreverence (Schächter and Taddeï 2010).

Technical innovations are generally not the result of sudden irruptions or unexpected inventions, but of the knowledge available at the time.

> The transformation of the technical milieu may find its conditions of possibility in the present imaginary, but which is already subject to recessions and actualizations of what is still virtual (Wunnenburger 2016).

The work of explaining and elucidating ideas that are still immature is accessible only to a modest number of people who are willing to invest themselves in exciting but risky operations. At an early stage, the complexity and multiplicity, even confusion, of the developments of any idea in the making become clear, with the need to draw out the essential lines in order to give it substance.

Apart from the survival of their little ivory tower, researchers increasingly live in an ever-changing business environment, in search of business productivity, performance and innovation, which implies internal changes (adaptation and adoption of new technologies, corporate culture changes) and external changes (customers, market, quality requirements, etc.). Disruptive innovations are one facet of these changes (Zubizarreta et al. 2021). In this sense, the ability to identify the technological disruptive potential of an invention at an early stage, coming from the academic world, could be a key factor for progress (Guo et al. 2019). However, we have, to put it simply and for different reasons, two worlds that are too disjointed and could be associated. Frameworks and models for ex ante identification of disruptive inventions are still few in number (Rafii and Kampas 2002; Keller and Hüsig 2009).

Instead of respecting Canguilhem's (1956) phrase, applied to 3D printing – "what a technique lacks is not the ability to discover the solution to its problems, but the ability to generalize these solutions" – because it was not considered, communication via various media was very useful in linking academic researchers with the economic world[11]. On the other hand, due to certain conceptual limitations, the other two additive manufacturing technologies have not been the subject of strong interactions with the business world. Table 3.2 shows the situation of 3D printing (3D, 4D, bio) between science and technology, located between scientific productions in their paradigmatic cocoons and economic realities (tension between continuity and discontinuity).

---

11 Available at: https://images.cnrs.fr/video/191.

Technique	3D	4D	Bio-printing
Science	+ Engineering science process materials	++ Promises	-- Epistemology
Technology; industry; economy	++ Strong development	- Expectations, except niches	High public expectation
	Science-Technology integration	'Free' science today	Interdisciplinary and convergent approaches not conducted

Table 3.2. *Positioning of 3D technologies in the conceptual space – applications*

Technological evolution is thus fuelled by different types of innovations. One of the most important types is disruptive innovation, which can have a significant impact on markets (Coccia 2018, 2020). This impact depends on previous – or even simultaneous – innovations, with a principle of competitive substitution (Coccia 2020a; Utterback et al. 2020). This dynamic competition between technologies can lead to the domination of a new technology over established technologies (Berg et al. 2019). With the emergence of 3D printing, the global market is estimated at 30€ billion/year, while the French market for mechanical engineering alone is around 80€ billion. We are not, as in electronics, at the stage of replacing lamps with transistors!

Nevertheless, 3D printing does not represent a major technological revolution, but it does represent a real breakthrough. Although with the other two technologies (4D and bio-printing), the only (exponential) development is in the number of scientific publications. In the case of 4D printing, there seems to be a missing link between Tibbits' (2013) expression of the idea and the somewhat chaotic scientific work being carried out (without questioning its scientific quality). For 3D printing of living matter, despite its enormous potential for applications (in the trillions of euros per year), a conceptual hurdle must be overcome, that of the transition from technology. Let us not forget that Vaucanson's historical approach was aimed at explaining the universe in mechanistic terms.

> This notion of model renders inoperative any clear-cut distinction between a purely 'artisanal' context of the automaton and another, properly philosophical, characteristic of rationalism (Wolf 2012).

3.4. References

Ancori, B. (2014). Communication, cognition et créativité dans les sociétés de la connaissance. À propos de deux dilemmes. *Nouvelles perspectives en sciences sociales*, 9, 45–93.

André, J.C. (2017a). *From Additive Manufacturing to 3D/4D Printing 1: From Concepts to Achievements*. ISTE Ltd, London, and John Wiley & Sons, New York.

André, J.C. (2017b). *From Additive Manufacturing to 3D/4D Printing 2: Current Techniques, Improvements and their Limitations*. ISTE Ltd, London, and John Wiley & Sons, New York.

André, J.C. (2017c). *From Additive Manufacturing to 3D/4D Printing 3: Breakthrough Innovations*. ISTE Ltd, London, and John Wiley & Sons, New York.

André, J.C. (2020). Impression 3D : niches applicatives porteuses. *Techniques de l'ingénieur*, BM7970 V2, 25.

André, J.C. (2021). Bio-printing. De l'organe à la médecine personnalisée, entre résultats et promesses. *Techniques de l'ingénieur*, RE268 v2.

Andreen, F. (2021). "Les Machiavéliens" de James Burnham, une lecture libérale [Online]. Available at: https://www.contrepoints.org/2022/07/22/408758-les-machiaveliens-de-james-burnham-une-lecture-liberale-i.

Ansart, P. (1990). Pluralisation des savoirs et formation scientifique. In *Pluralité des enseignements en sciences humaines à l'université*, Racette, G. and Forest, L. (eds). Noir sur Blanc, Montreal.

Arasse, D. (2000). *On n'y voit rien*. Denoël, Paris.

Arpin, I., Barreteau, O., Caranta, C., Ducrot, C., Garin, P., Hannachi, M., Maillet, I. (2020). Séminaire d'échanges sur les pratiques de recherches interdisciplinaires à INRAE [Online]. Available at: https://hal.inrae.fr/hal-02890162v2/file/2020_Juill%2010__ValorisationS%C3%A9minaireInterdisciplinarit%C3%A9.pdf.

Arrow, K.J. (1951). Alternative approaches to the theory of choice in risk-taking situation. *Econometrica*, 17, 404–437.

Augé, M. (1992). *Non lieux, introduction à une anthropologie de la sur-modernité*. Le Seuil, Paris.

Avenier, M.J. (2019). Les Sciences de l'artificiel : une conceptualisation révolutionnaire de sciences fondamentales à parachever. *Projectics/Proyéctica/Projectique*, 24, 43–56.

Bachelard, G. (2003). *Le nouvel esprit scientifique*. Puf, Paris.

Bachimont, B. (2005). Ingénierie des connaissances, ingénierie de la contingence : la technique entre le nécessaire et le singulier. In *Entre connaissance et organisation : l'activité collective*, Lorino, P. and Teulier, R. (eds). La Découverte, Paris.

Barthélémy, J.H. (2013). Encyclopédisme et théorie de l'interdisciplinarité. *La Presse Hermès*, 67, 165–170.

Bazerman, M.H. and Moore, D.A. (2009). Introduction to managerial decision making. In *Judgment in Managerial Decision Making*, Bazerman, M.H. and Moore, D.A. (eds). Wiley, New York.

Berg, S., Wustmans, M., Bröring, S. (2019). Identifying first signals of emerging dominance in a technological innovation system: A novel approach based on patents. *Technological Forecasting and Social Change*, 146, 706–722.

Berger, G. (1972). Opinions et réalités. In *L'interdisciplinarité. Problèmes d'enseignement et de recherche dans les universités*, Apostel, L., Berger, G., Briggs, A., Michaud, G. (eds). Organisation de coopération et de développement économiques, Paris.

Bernard, C. (1965). Leçons sur les phénomènes de la vie communs aux végétaux et aux animaux [Online]. Available at: http://claudebernard.in2p3.fr/index.php?num_page=92.

Bessant, J., Moslem, K., Neyer, A.N., Piller, E., Von Stamm, B. (2010). Radical innovation: Making the right bets [Online]. Available at: https://ore.exeter.ac.uk/repository/handle/10871/14934.

Bibard, L. (2017). Pour un avenir réellement collaboratif : les écueils à éviter. In *Réinventer le leadership*, Frimousse, S. and Le Bihan, Y. (eds). EMS, Paris.

Bienaise, J. (2016). D'une interdisciplinarité-processus à une interdisciplinarité-œuvre. *L'Annuaire théâtral*, 60, 91–103.

Boffa-Comby, P. (2019). Il suffit de 10% des collaborateurs pour changer toute l'entreprise [Online]. Available at: https://www.hbrfrance.fr/chroniques-experts/2019/09/27890-il-suffit-de-10-des-collaborateurs-pour-changer-toute-lentreprise/.

Bognon-Küss, C. (2012). Le vitalisme est-il un indéterminisme ? In *Le déterminisme entre sciences et philosophie*, Charbonnat, P. and Pépin, F. (eds). Éditions Matériologiques, Paris.

Bonnardel, N. and Lubart, T.I. (2019). La créativité : approches et méthodes en psychologie et en ergonomie. *Revue interdisciplinaire management, homme & entreprise*, 37, 79–98.

Boukamel, O. and Emery, Y. (2018). Les freins culturels à l'innovation dans l'administration publique : spécificités helvétiques. *Gestion et management public*, 6, 25–43.

Bouveresse, J. (2021). Intellectuels médiatiques et penseurs de l'ombre [Online]. Available at: https://www.monde-diplomatique.fr/2006/05/BOUVERESSE/13428.

Brownlie, S. (2008). Resistance and non-resistance to boundary crossing in translation research. *Target*, 20, 333–347.

Burnham, J. (1964). *The Machiavellians: The Defenders of Freedom*. Gateway, New York.

Bush, V. (1945). The endless frontier. National Science Foundation [Online]. Available at: https://www.nsf.gov/about/history/nsf50/vbush1945_content.jsp.

Canguilhem, G. (1956). La pensée de R. Leriche. *Revue philosophique de la France et de l'étranger*, 146, 313–317.

Canguilhem, G. (1957). La formation du concept de réflexe aux XVIIe et XVIIIe siècles. *Revue de métaphysique et de morale*, 62, 99–101.

Casakin, H.P. (2007). Metaphors in design problem solving: Implications for creativity. *International Journal of Design*, 1, 23–35.

Chauvey, J.N. and Georgescu, I. (2017). Les heuristiques du jugement en contrôle de gestion : le cas des décisions rapides. Responsabilités et comptabilités [Online]. Available at: https://hal.archives-ouvertes.fr/hal-01907562/document.

Cholle, F. (2020). *Squircle, a New Way to Think for a New World*. Squircle Academy, New York.

Classens, M. (ed.) (2016). Science et non-décision. In *Décider de ne pas décider*. Quae, Versailles.

Coccia, M. (2018). Disruptive firms and technological change. *Quaderni*, 3, 3–18.

Coccia, M. (2020a). The evolution of scientific disciplines in applied sciences: Dynamics and empirical properties of experimental physics. *Scientometrics*, 124, 1–37.

Coccia, M. (2020b). Asymmetry of the technological cycle of disruptive innovations. *Technology Analysis & Strategic Management*, 32, 1462–1477.

Collinet, C. and Terral, P. (2014). Le travail scientifique interdisciplinaire : réalités et contraintes [Online]. Available at: https://www.parisnanterre.fr/medias/fichier/rapport-trascinter-version-finale_1412093003347-pdf.

Cropley, D.H. (2015). *Creativity in Engineering. Novel Solutions to Complex Problems*. Academic Press, London.

Csikszentmihalyi, M. (2006). *La créativité*. Robert Laffont, Paris.

Danzo, A. and André, J.C. (2020). Que peut apporter une modélisation mathématique à la maîtrise du bio-printing ? *Entropie*, 20(1), 32 [Online]. Available at: http://openscience.fr/Que-peut-apporter-une-modelisation-mathematique-a-la-maitrise-du-bio-printing.

De Casabianca, D. (2019). Ne pas déranger, idées en gestation. In *Gaston, un philosophe au travail*, Ortoli, G. (ed.). Philosophie Magazine, Paris.

Deleuze, G. and Guattari, F. (1972). *Capitalisme et schizophrénie*. Éditions de Minuit, Paris.

Demoly, F. and André, J.C. (2021a). Impression 4D : une cécité de confort et/ou une approche systé-mique empêchée ? *Entropie*, 49 [Online]. Available at: https://www.openscience.fr/IMG/pdf/iste_entropie21v2n1_1.pdf.

Demoly, F. and André, J.C. (2021b). Impression 4D : promesses ou futur opérationnel ? *Techniques de l'ingénieur*, RE285 v1.

Demoly, F. and André, J.C. (2022a). Roadmapping 4D printing through disruptive ideas. In *Smart Materials in Additive Manufacturing, 1: 4D Printing Principles and Fabrication*, Bodaghi, M. and Zolfagharian, A. (eds). Elsevier, Amsterdam.

Demoly, F. and André, J.C. (2022b). *4D Printing 1: Between Disruptive Research and Industrial Applications*. ISTE Ltd, London, and John Wiley & Sons, New York.

Descartes, R. (1996). Le modèle du vivant dans la physique cartésienne. *Philosophiques*, 23, 389–404.

Detienne, M. and Vernant, J.P. (1974). *Les ruses de l'intelligence. La métis des Grecs*. Flammarion, Paris.

Drucker, P. (2006). *Innovation and entrepreneurship*. Collins, New York.

Dumont, N.T. (2018). Les heuristiques fonctionnelles [Online]. Available at: https://www.westattitude.com/single-post/2018/04/25/LES-HEURISTIQUES-fonctionnelles.

Dumora, F. (2020). De l'homme renaissant aux humanités numériques. Littérature, culture et société : construire une pensée interdisciplinaire [Online]. Available at: https://hal.univ-reims.fr/tel-03231332/document.

Einstein, A. (2007). *L'espoir économique : vers la révolution du quaternaire*. Bourin, Paris.

Eisenhower, D.D. (2018). Décider dans l'incertitude [Online]. Available at: http://tsm-alumni.fr/files/ALUMNI/MRH/andrh/Compte-Rendu-ANDRH-8-Fevrier-2018.pdf.

Elgozy, G. (1966). *Le paradoxe des technocrates*. Denoël, Paris.

Feyerabend, P. (1988). *Contre la méthode ; esquisse d'une théorie anarchiste de la connaissance*. Le Seuil, Paris.

Findling, C., Skvortsova, V., Dromnelle, R., Palminteri, S., Wyart, V. (2019). Computational noise in reward-guided learning drives behavioral variability in volatile environments. *Nature Neuroscience*, 22, 2066–2077.

Foucart, J. (2008). Travail social et construction scientifique. *Pensée plurielle*, 19, 95–103.

Fourez, G. (2001). *La construction des sciences. Les logiques des inventions scientifiques : introduction à la philosophie et à l'éthique des sciences*. De Boeck, Brussels.

Galano, J.M. (2020). Transversalité et conflits. *XMP Consult*, 6, 4–5.

Gambier, Y. (2019). Traductologie et médias : les défis de l'interdisciplinarité. *Interdisciplinarity and Translation Studies*, 9–10, 11–24.

Ganascia, J.G. (1999). *2001, l'odyssée de l'esprit*. Flammarion, Paris.

Ge, Q., Sakhaei, A.H., Lee, H., Dunn, C.K., Fang, N.X., Dunn, M.L. (2016). Multi-material 4D printing with tailorable shape memory polymers. *Scientific Reports*, 6, 31110.

Gigerenzer, G. and Gaissmaier, W. (2011). Heuristic decision making. *Annual Review of Psychology*, 62, 451–482.

Giget, M. (2020). Importance de la vision prospective pour les directions de l'innovation [Online]. Available at: https://www.youtube.com/watch?v=WayIBS-IDDI.

Golse, B. (2004). La naissance des représentations. Conceptions psychanalytiques. *Nouveau traité de psychiatrie de l'enfant et de l'adolescent*, 173–188.

Grieu, E. (2006). "Le désir et la promesse" CERAS "risques partagés" [Online]. Available at: http://www.ceras-projet.com/lodel/documents.php?id=1951&format=print.

Grimaud, E., Tastevin, Y.P., Vidal, D. (2017). Low tech, high tech, wild tech. Réinventer la technologie ? *Techniques & Culture*, 2017/1, 12–29.

Guo, J., Pan, J., Guo, J., Gu, F., Kuusisto, J. (2019). Measurement framework for assessing disruptive innovations. *Technology Forecasting Social Change*, 139, 250–265.

Habermas, J. (2002). *L'avenir de la nature humaine ; vers un eugénisme libéral ?* Gallimard, Paris.

Hahn, F., Jensen, S., Tanev, S. (2014). Disruptive innovation vs disruptive technology: The disruptive potential of the value propositions of 3D printing technology startups. *Technology Innovation Management Review*, 4, 27–36.

Hanaphy, P. (2022). Scientists say India's first 3D printed cornea "can be used in humans" after successful animal trial [Online]. Available at: https://3dprintingindustry.com/news/scientists-say-indias-first-3d-printed-cornea-can-be-used-in-humans-after-successful-animal-trial-213745/.

Haroche, J.C. (2006). Effets psychiques du mouvement continu et de l'illimitation dans l'exercice contemporain de la pensée. In *Désir de penser, peur de penser : situations critiques*, Enriquez, E., Haroche, C., Spurk, J. (eds). Paragon, Lyon.

Hatano-Chalvidan, M. (2010). *Dires sur l'interdisciplinarité : la référence à l'interdisciplinarité dans le discours des chercheurs en sciences exactes*. Éditions universitaires européennes, Sarrebruck.

Hempel, C. (1966). *Philosophy of Natural Science*. Prentice Hall, New York.

Horenstein, M.N. (2002). *Design Concepts for Engineers*. Prentice Hall, New York.

James, P. (2011). Symposium on interdisciplinary approaches to international studies: History, psychology, technology studies, and neuroeconomics. *International Studies Perspectives*, 12, 89–93.

Jurdant, B. (2009). Communication scientifique et réflexivité. *Science & Société* [Online]. Available at: https://science-societe.fr/baudouin-jurdant-intervention-communication-scientifique-et-reflexivite-2009/ [Accessed 7 November 2023].

Keller, A. and Hüsig, S. (2009). Ex ante identification of disruptive innovations in the software industry applied to web applications: The case of Microsoft's vs. Google's office applications. *Technology Forecasting Social Change*, 76, 1044–1054.

Keynes, J.M. (2016). La prospective : de l'anticipation à la préparation à l'avenir. *Réalités industrielles – Annales des Mines*, 33–37.

Klein, J.T. (2001). Interdisciplinarity and the prospect of complexity: The tests of theory. *Issues in Integrative Studies* 19, 43–57.

Kobarg, S., Stumpf-Wollersheim, J., Welpe, I.M. (2019). More is not always better: Effects of collaboration breadth and depth on radical and incremental innovation performance at the project level. *Research Policy*, 48, 1–10.

Kourilsky, F. (2002). Le chemin de l'interdisciplinarité. In *Ingénierie de l'interdisciplinarité*, Kourilsky, F. (ed.). L'Harmattan, Paris.

Kozbelt, A., Beghetto, R.A., Runco, M.A. (2010). Theories of creativity. In *Cambridge Handbook of Creativity*, Kaufman, J.C. and Sternberg, R.J. (eds). Cambridge University Press, Cambridge.

Lambert, F. (2015). La place du nom erreur et du verbe se tromper dans la langue française et la fonction heuristique de l'erreur dans la recherche. *Essais*, 8, 18–25.

Lavelle, L. (1992). *De l'acte*. Aubier, Paris.

Le Comptoir (2018). Retour sur un phénomène en accélération : la disruption [Online]. Available at: https://lecomptoir-cpr.fr/Common-Content/Actualites/France/2017/Experts/RETOUR-SUR-UN-PHENOMENE-EN-ACCELERATION-LA-DISRUPTION.

Le Fur, J. (2014). Du foisonnement des disciplines à la recomposition d'une réalité partagée : élaboration d'une structure de modélisation dédiée à l'intégration de connaissances disciplinaires. In *XXIème journées de Rochebrune "Multi-trans-interdisciplinarité". Rencontres interdisciplinaires sur les systèmes complexes naturels et artificiels. Rochebrune* [Online]. Available at: https://horizon.documentation.ird.fr/exl-doc/pleins_textes/2022-02/010084143.pdf.

Lebeau, A. (ed.) (1996). Choix techniques et décision politique. In *Colloque de la Villette "Le savant et le politique aujourd'hui : un débat sur la responsabilité et la décision"*. Albin Michel, Paris.

Lenoir, Y. (1995). L'interdisciplinarité : aperçu historique de la genèse d'un concept. *Cahiers de la recherche en éducation*, 2, 237–265.

Lesage, M.C. (2008). L'inter-artistique : une dynamique de la complexité. *Registres*, 13, 11–26.

Lévi-Strauss, C. (1990). *La pensée sauvage*. Le livre de poche, Paris.

Lévi-Strauss, C. (ed.) (2020). Diogène couché. In *Lévi-Strauss : une anthropologie sans philosophie ?* Puf, Paris.

Lévy-Leblond, J.M. (2006). *La vitesse de l'ombre ; aux limites de la science*. Le Seuil, Paris.

Livio, M. (2017). *Fabuleuses erreurs. De Darwin à Einstein*. CNRS, Paris.

Lizarralde, I. and Pilnière, V. (2012). Les phases amont des projets innovants et la construction du groupe-projet. Pour une approche socio-cognitive. *Projectics/Proyéctica/Projectique*, 10, 97–110.

Lizarralde, I., Larrasquet, J.M., Coutts, N. (2011). Design and innovation in the face of complexity (Towards new challenges of linking systems and learning). *Projectics/Proyéctica/Projectique*, 8–9, 199–211.

Louargant, S. (2003). De l'interdisciplinarité proclamée à la multidisciplinarité évoquée : le territoire comme enjeu des territoires disciplinaires. Une lecture transversale [Online]. Available at: https://halshs.archives-ouvertes.fr/halshs-00261774/document.

Lubart, T. (2010). *Psychologie de la créativité.* Armand Colin, Paris.

Mazurek, H. (2016). Le partage des savoirs. Un nouveau paradigme en sciences sociales ? [Online]. Available at: https://horizon.documentation.ird.fr/exl-doc/pleins_textes/divers16-2/010066392.pdf.

McCauley, M. (2012). Creative thinking: Linking environment, vision, change, and strategy. *Small War Journal* [Online]. Available at: http://smallwarsjournal.com/jrnl/art/creative-thinking-linking-environmentvision-change-and-strategy.

McMurtry, A. (2009). Knowers and phenomena: Two different approaches to interdisciplinarity and inter-professionalism. *Issues in Integrative Studies*, 27, 1–16.

Miailhe, D., Balas-Ezzemzami, J., Combes, J.M., Lojou-Julien, C. (2017). Penser la complexité, un étayage pour la pratique managériale. *Empan.*, 107, 81–87.

Michels, R. (1914). *Les partis politiques ; essai sur les tendances oligarchiques des démocraties*. Flammarion, Paris.

Millier, P. (2016). L'ingénieur, le bricoleur et l'innovateur. *Entreprendre & innover*, 28, 48–62.

Mishima, Y. (1989). *Spring Snow*. Gallimard, Paris.

Montgomery, N., Squires, G., Syed, I. (2018). Disruptive potential of real estate crowdfunding in the real estate project finance industry. A literature review. *Property Management*, 36, 597–619.

Morin, E. (1998). Réformons la pensée. *Le monde de l'éducation*, 252, 28.

Morin, E. (2006). *La méthode.* Le Seuil, Paris.

Muller, R.C. (1993). Enhancing creativity, innovation and cooperation. *AI & Society*, 7, 4–39.

Mumford, M.D., Todd, E.M., Higgs, C., McIntosh, T. (2017). Cognitive skills and leadership performance: The nine critical skills. *The Leadership Quarterly*, 28, 24–39.

Neumann, I.B. (2011). End comment: The practice of interdisciplinarity. In *International Studies. Interdisciplinary Approaches*, Aalto, P., Harle, V., Moisio, S. (eds). Macmillan, New York.

Newell, W.H. (2001). A theory of interdisciplinary studies. *Issues in Integrative Studies*, 19, 1–25.

O'Reilly, C. and Binns, A.J.M. (2019). The three stages of disruptive innovation: Idea generation, incubation and scaling. *California Management Review*, 61, 49–71.

Pareto, V. (1935). *L'esprit et la société*. Harcourt, Brace and Co., New York.

Parrochia, D. (1994). Le statut épistémologique de la vie artificielle. In *Ordre biologique. Ordre technologique*, Tinland, F. (ed.). Champ Vallon, Paris.

Patton, G. (2017). À la découverte de la disruption [Online]. Available at: https://leblogdudisrupteur.com/2017/04/03/cest-quoi-la-disruption-1/.

Paty, M. (1990). *L'analyse critique des sciences ou le tétraèdre épistémologique*. L'Harmattan, Paris.

Pomerol, J.C. (2012). *Decision Making and Action*. ISTE Ltd, London, and John Wiley & Sons, New York.

Poutanen, P., Soliman, W., Ståhle, P. (2016). The complexity of innovation: An assessment and review of the complexity perspective. *European Journal of Innovation Management*, 19, 189–213.

Rafii, F. and Kampas, P.J. (2002). How to identify your enemies before they destroy you. *Harvard Business Review*, 80, 115–134.

Repko, A.F. (2012). *Interdisciplinary Research: Process and Theory*. Sage, London.

Repko, A.F., Newell, W.H., Szostak, R. (2012). *Case Studies in Interdisciplinary Research*. Sage, London.

Resweber, J.P. (1981). *La méthode interdisciplinaire*. Puf, Paris.

Revel, J.F. (1988). *La connaissance inutile*. Grasset, Paris.

Rey, A. (1992). *Dictionnaire historique de la langue française*. Le Robert, Paris.

Reynaud, J.D. (1997). *Les règles du jeu. L'action collective et la régulation sociale*. Armand Colin, Paris.

Rose, J. (2001). *Profession quasi-chercheur*. L'Harmattan, Paris.

Rosset, C. (1984). *Le réel et son double*. Folio Essais, Paris.

Samsara (2016). Primary human hepatic stellate cells [Online]. Available at: http://www.samsarasciences.com/.

Schächter, V. and Taddei, F. (2010). La créativité dans la recherche. *Journal de l'école de Paris du management*, 83, 29–36.

Schütz, A. (2008). *Le chercheur et le quotidien. Phénoménologie des sciences sociales*. Klincksieck, Paris.

Sérieyx, H. (2018). La confiance, composante vitale de la performance dans un monde nouveau. In *Trusting Organizations*, Sérieyx, H. and Riendeau, D. (eds). EMS, Paris.

Sevtsuk, A., Chancey, B., Basu, R., Mazzarello, M. (2022). Spatial structure of workplace and communication between colleagues: A study of E-mail exchange and spatial relatedness on the MIT campus. *Social Networks*, 70, 295–305.

Silberzahn, P. and Rousset, B. (2019). *Stratégie modèle mental : cracker enfin le code des organisations pour les remettre en mouvement*. Diateino, Paris.

Sinacœur, M.A. (1983). Qu'est-ce que l'interdisciplinarité ? In *Interdisciplinarité et sciences humaines*, Apostel, L., Benoist, J.M., Bottomore, T.B., Boulding, K.E., Dufrenne, M., Eliade, M., Furtado, C., Gusdorf, G., Krishna, D., Mommsen, W.J. et al. (eds). UNESCO, Paris.

Sinacœur, M.A. (ed.) (1992). Quelques réflexions sur l'interdisciplinarité. In *Entre savoirs : l'interdisciplinarité en acte ; enjeux, obstacles, résultats*. Erès, Toulouse.

Soulet, M.H. (2017). Les épreuves de l'incertain [Online]. Available at: https://calenda.org/404235?file=1.

Stengers, I. and Prigogine, I. (1979). *La nouvelle alliance*. Gallimard, Paris.

Tan, D., Nokhodchi, A., Maniruzzaman, M. (2019). 3D and 4D printing technologies: Innovative process engineering and smart additive manufacturing. In *3D and 4D Printing in Biomedical Applications: Process Engineering and Additive Manufacturing*, Maniruzzaman, M. (ed.). Wiley-VCH, Verlag, Berlin.

Thomas, C., Tibbits, S., Banning, G. (2014). The programmable world. *Scientific American*, 311, 60–65.

Tibbits, S. (2014). 4D printing: Multi-material shape change. *Architectural Design*, 84, 116–121.

Tibbits, S. (2016). *Self-Assembly Lab: Experiments in Programming Matter*. Routledge, New York.

Tibbits, S. and Tomas, A.F. (2013). Biomolecular, chiral and irregular self-assemblies. *Acadia Adaptative Architecture*, 261–268 [Online]. Available at: http://papers.cumincad.org/data/works/att/acadia13_261.content.pdf.

Turing, A. (1952). The chemical basis of morphogenesis. *Philosophical Transactions of the Royal Society of London*, 237, 37–72.

Utterback, J.M., Pistorius, C., Yilmaz, E. (2020). The dynamics of competition and of the diffusion of innovations. MIT Sloan School of Management Working Paper, 6054-20 [Online]. Available at: https://hdl.handle.net/1721.1/124369.

Vessier, L. (2022). Deux enquêtes dressent un portrait contrasté de l'état du doctorat en France [Online]. Available at: https://themeta.news/les-deux-visages-du-doctorat/.

Vian, B. (1956). *L'automne à Pékin*. 10/18, Paris.

Walras, L. (1926). *Éléments d'économie politique pure*. Pichon/Durand-Auzias, Paris.

Weil, S. (1991). *Gravity and Grace*. Agora, Paris.

Weil, T. and Dubey, A.S. (2020). Transformer l'entreprise autour d'un projet partagé. *The Conversation* [Online]. Available at: https://theconversation.com/transformer-lentreprise-autour-dun-projet-partage-130887.

Wolf, C.T. (2012). Le mécanique face au vivant. In *L'automate : modèle, machine, merveille : la philosophie de la biologie avant la biologie. Une histoire du vitalisme*, Gaillard, A., Roukhomovsky, B., Roux, S., Le Goffi, J.Y. (eds). Garnier, Paris.

Wu, L., Wang, D., Evans, J.A. (2019). Large teams develop and small teams disrupt science and technology. *Nature*, 566, 378–382.

Wunnenburger, J.J. (2016). Explorations d'imaginaires. In *L'imaginaire*, Wunnenburger, J.J. (ed.). Que sais-je, Paris.

Zola, E. (1997). *The Ladies' Paradise*. Le livre de poche, Paris.

Zubizarreta, M., Ganzarain, J., Cuadrado, J., Lizarralde, R. (2021). Evaluating disruptive innovation project management capabilities. *Sustainability*, 13, 1.

Conclusion

> We rely on technical progress to help us overcome the problems we encounter and know to be real – practices and attitudes from which we have not yet emerged.
>
> (Leroux 2011)

> Doesn't risk ultimately appear to be the impoverished and reductive way in which the Man of technoscientific societies, who can no longer make sense of his misfortune, accounts for what happens to him.
>
> (Dupuy 2002)

> If our cells are free, their organization must obey the principles of economy or ecology that structure societies or ecosystems. In an ecosystem such as the forest as a whole, there are general regulatory phenomena that control temperature or soil composition. However, there is no such thing as a forest genome that, by programming each of its inhabitants, ensures the realization of global phenomena. Homeostasis – the constancy of the forest's internal environment – is not written down anywhere. There is no nature police.
>
> (Sonigo 2000)

The management of research, the project-based approach, assessment of interdisciplinary projects, the position of academic research in relation to the business world, the ability to accept responsibility and ethics are some of the questions raised in this book, along with the somewhat hazardous and difficult link between idea and innovation; this is a key element in the development of promising emerging technologies. To conclude this reflection on creativity, invention and innovation, from someone with an academic background, the author has "amused" himself by producing a short tale on his vision of research, as presented below.

C.1. A little tale of (almost) the end

The bone is broken, and I gather (from what I am told) that it is complicated. So, I imagine a process in which biological microsystems revised by chemists and engineers will produce bone. I am also realizing that blood circulation is not much use as a transport route, because of the dilution of the bio-vector, so it has to be introduced close to the injured area, located and transported to that area, and fragmented into the active part that will be deposited on the area being repaired. Aside from biology, of which I know little, for such a project, I can contribute some knowledge (from vague to in-depth) of fluid mechanics, wave–matter interactions, chemistry, and process engineering. In short, all of the disciplines that are explored and exploited in "additive manufacturing". We can no doubt imagine just as many, if not more, on the bio side, not to mention medical constraints. It is a good idea to have this accident to think about... but my coffee is cold.

I told myself that when I went "to the French capital", I would not see the Eiffel Tower, but the "temple of science". They must know how to do it, or simply explain to me how to link these disciplines. When I got to the temple, they let me in. Indeed, inside, seeing the number of people passing each other with files under their arms, stamped "TS", you can imagine the intense activity that takes place in its various premises! But no, I am told, science is spread all over France. The "temple of science" has offices for its priests (those with files under their arm?), analyzes scientific data from its laboratories with its machines and keeps its tables of law. These are based on concepts such as scientific discipline, excellence, evaluation, international influence, and it is precisely with all this global data that the body of doctrine can evolve incrementally (but, apart from disruptions, only marginally). With all this, we come up with rankings and balance sheets. In the end, would we have done research just to come up with cost accounting? All those buildings for just one figure! All those people at work (usually more than 35 hours a week) for that and a few trinkets and other odds and ends for communication.

At the 1933 World's Fair in Chicago, it was written: "Science finds, industry applies, man conforms" (Encyclopedia of Chicago 1934). I get the impression that, with industrial contracts, it is industry that innovates (invents) and even that it is the public that becomes concerned by creativity and innovation (making a useful "object"). However, in 1933, we associated science, invention and innovation. This was already possible because the world was moving at a slower pace, but with major scientific discoveries.

So, today, we have translated the exciting scientific activity that produces radically new knowledge, a factor of real technological progress, into the activity of a shopkeeper or an accountant to make "notons" (superficial administrative certificates), as with the French author Boris Vian, finally, to respond to the reductionist need of the temple. This scientific work is separated from invention, and even more so from innovation! But then, how can we accept situations where current scientific performance and inventions are so strongly separated? Quite simply by having two systems, one associated with science and the other with objects. Depending on the importance we give to these two components, we will be innovative or not. It is important to remember that the successes in other countries show that you have to trust the jacks-of-all-trades and the DIYers. It is a bit like President Macron's idea with his startup nation promotion. But, we are far from science!

So, I asked, is a DIYer (scientist) only supposed to be competent if they can keep up with the evolution of the sciences they need to create? As these evolve at least quantitatively quickly, their creativity based on what is real and what is possible is likely to diminish over time, whereas a disciplinary scientist or a temple priest has the time of a career. In spite of this, a small number of people have projects that go beyond their discipline and are supported. In the current, relatively percolating structure, this is a way of stimulating projects outside the discipline, so that the bearers can be reintegrated into their (or any other) discipline. However, we are not quite there yet and we are forgetting a very important element, namely that this science-object goes from concept development to the realization of a demonstrator.

Browsing a synthetic chemistry journal recently, I noticed that in most of the articles, the researchers were working on very small quantities of matter. With Fred (we have been cooperating in a spirit of friendship and trust for a few "long weeks"), we needed a chemical substance with certain properties for a POC. The author of the work on a substance with these properties was unable to supply us with the required quantity quickly (not mg, but more than a thousand times more), and was unwilling to waste his time making it, which was not part of his mission. In the circumstances, we were forced to find an equivalent (or close equivalent) on the "shelf"! It is a translation of the DIY spirit mentioned above that is presented here. If you can achieve your goal with a little imagination, you can move fast enough to modify the system and present your prototype on time. With the ability to experiment on this device, progress can then be made.

With regards to the initial problem at hand, it is not possible to have a research structure adapted to each subject, especially if it requires bringing together a large number of disciplines that do not know each other, and do not talk to each other (because they are spatially separated and/or do not use the same language when exploring their own paradigms). A hopeless situation! Such a project represents too great a financial risk, especially if exchanges between communities do not take place properly. And yet, the organization is accountable, and in the end, governance corresponds to the search for the least risk for scientific work within their paradigm, which guarantees the best performance according to the financial criteria of the temple of science.

What about innovation? It is largely outsourced, leaving most of it in the hands of industry, with different sources of funding. There can be particularly fruitful exchanges, but also simple "customer-supplier" relationships. France suffers from a composite industrial fabric, but not at the best level compared to Germany: globalized companies, a few large companies with strong links to research, etc. All this is well known.

Does this mean that inventions only come from industry? A large majority of patents come from industry, but TS still holds a strong position on a national scale. Fred and I "experimented" with filing a patent via TS, and with the various "aids" we received, we lost more than three years! So, by streamlining support, it would not cost the system too much, or even less (and it would be more accountable), and we would gain in efficiency. Incidentally, I looked at some of the patents in our field, and those who held them, on the same invention, were often from different disciplinary backgrounds (see the "bricolage savant" operations where disjointed knowledge can be exploited).

Interdisciplinarity requires translation and mediation services related to intellectual issues that go far beyond science. In my opinion, they are philosophical. They have to do with the meaning of knowledge, all knowledge, considered as the experience of the human species, aligned with its will to know and its biological survival (Hamel 2002).

Tell yourself *a priori* that something is going to change. We set ourselves a goal, we launch propositions, we sow various seeds everywhere, and we know that time will bring an answer (Moreno 1990).

Who in the scientific world has ever thought of advocating and endorsing in research activities the values of distance and detour (forcing oneself to look elsewhere for new leads), the values of incompetence (enlisting the cooperation of people who know nothing about the field being prospected) or the values of insolence (favoring mental travelling, the disconcerting, even the absurd)? (Besnier 2013).

The social Man is himself fragmented according to his economic, political and cultural activities and beliefs; according to his space or history, and so on. Specialized knowledge communicates only inadequately with one another, each remaining locked in its own language. Fortunately, there are some undisciplined minds who move from one discipline to another. But these transgressions remain individual. Hubert Curien used to say that, generally speaking, scientists are like wolves: they urinate to mark their territory and bite any intruder who enters (Morin 1999).

Absolute cooperation, obedience and support will increase the availability of sources of power, and thus the capacity of rulers to exercise power. Conversely, restricting or refusing cooperation will reduce or eliminate the availability of these sources of power. The fact of cooperating or not opens or closes access to the resources needed by the ruler, just as a faucet controls the flow of water. [...] If, then, as La Boétie concluded, "tyrants are given nothing, they are obeyed nothing, without a fight, without a blow, they find themselves naked and defeated, unable to go any further, just as the root deprived of soil or food lets the branch wither and die (Sharp 1995).

In an ideal framework, an interdisciplinary decision-making and functional interdisciplinarity should bring together various disciplinary viewpoints to build a kind of theoretical mesh around the complex problem. The more complete this edifice of information is, the more it will be able to "make the 'real' complex problem speak". A truly collective approach is therefore necessary (Naudon 2013).

Interdisciplinarity will always remain a necessary but insufficient means of restoring a global approach to problems. Separation between disciplines can only help technocratic and bureaucratic powers to pursue their dominant strategies, if there is disciplinary compartmentalization. In this sense, interdisciplinarity is, in a university and societal context, a democratic challenge par excellence (Thill 1987).

A structure is an operational set of indefinite meaning [...] grouping elements, in any number, whose content is not specified, and relations in finite number, whose nature is not specified, but whose function is defined (Serres 1969).

Box C.1. *A few phrases on and around interdisciplinarity*

The word "interdisciplinarity" emerges belatedly. In short, it is a way of giving a certain social recognition to this statement of action, but little more. Following the example of biophysics, an active critical mass of "biophysicists" fought for the creation of a new discipline. It was then integrated into the group of disciplines in the temple of science accepting its rules of the game. In fact, by removing the "R" from scientific revolution, a method of incremental enrichment for the temple is translated into a new discipline that gains recognition (jobs, finances) by fitting into its "authorized" paradigm.

This situation, as with patents, illustrates the life of the basic researcher who steps out of the "theoretical framework" (we still have, and all claim, the right to think in TS) to exchange with other researchers, tinker together on subjects at the interfaces, and without seeking to create a new discipline, to be financially supported for such bi-disciplinary operations (or operations involving even more than two disciplines). And before the mission for interdisciplinarity (MI), it was not easy to develop a joint project without internal support. Today, in the context of calls for projects, for modest sums of money and a low success rate, there are ways of obtaining funding. But out of mimicry or methodological conservatism, we end up with the same project selection methods, with results that are sometimes “astonishing”. In short, why not, but it is a bit of a stretch.

I believe that the MI is as much a part of the temple of science as any other discipline. This undoubtedly means recognizing its social usefulness, without giving it the means to fulfill the many scientific and technological ambitions that it should be supporting, guiding, enriching and retaining. In short, open and living science! So, when is the revolution to create a real space for freedom and/or action, where impertinent initiatives can be supported responsibly? But this mad hope may never come true, as researchers feel less and less concerned by TS, just for a salary and a few subsidies, with evolving but ever-widening relational circles.

What will the temple be used for in the future? Everything is moving around us, even politics (at least on the surface), but the scientific world still seems stable because everyone has developed their own circle of relationships. So, for TS, as in *The Leopard*: “If we want things to stay as they are, things will have to change”. Apart from keeping up appearances, as in many other places, it is a loss of connection between the rank and file, and the hierarchy. It is a sign of the decline in the system’s “good” governance and its detachment from society’s real needs. And there are so many of them.

Thank you very much for having given me so much of your time and so much information, even if you describe a future that is less and less adapted to today’s world (and to the future) and who does not know that the temple and its disciplines is going to explode, but that is probably not very serious. To thank you for your kindness, would you allow me to offer you a modest present that is a product of my campaign? Simply write on the envelope “Mr. Guardian of the Temple of Science, it will happen!” And again, little has been said about Cowen:

> We can’t understand why we can’t anymore. All these problems have a single, unnoticed cause: for at least three hundred years, we’ve been living off fruit that was just waiting to be picked [...]. But over the past forty years, these fruits have become scarce, and we’ve acted as if they were still there. We didn’t want to recognize that we had reached a technological plateau and that the tree was much barer than we wanted to admit. So, as soon as a guideline seems to bring satisfactory technical progress technical progress, we run the risk of continuing to pursue creative action through incremental research, seeking to keep on winning without making a clean break. But this quest clearly has its limits (Cowen 2011)[1].

1 Author’s translation from the French version of the text.

C.2. Conclusion

Since time immemorial, man-made objects have been valued as multipliers, facilitators of human effort. Tools and machines are extensions of the hand, and the source of power and the desire to increase it by transforming ideas into practical achievements. The consumer society has exacerbated this desire to have more and more socialized novelties to offer, and to expect creative people to come up with the disposable and ephemeral pursuit of their dreams! "It took a century, from 1870 to 1970, for gross world product (GWP) to rise from $2,000 billion to $15,000 billion. But since 1970, in just fifty years, this GWP has risen from $15,000 billion to $84,000 billion" (Trégouët 2020).

In today's increasingly challenging economic and competitive environment, radical ideas are a must if companies are to remain competitive. The very early stages of the invention process are still fragile and need to be clarified (and supported). This process, at the very beginning of the idea-application chain, is difficult but essential to lay the foundations for possible success, right through to the emergence of a new, equally profitable idea. One of the key skills is to identify and develop all of the possibilities in these early stages. Once ideas have been identified as realistic, they then need to be framed, a POC developed and supported to the extent that they can reach the company in a market, until they can be integrated for a time into the product development process. Selecting radically new ideas is neither easy nor intuitive. What is more, it can be conservatively difficult to maintain their support throughout the development process (Nicholas et al. 2015; Colombo et al. 2017).

While the real markets for industrial companies are now globalized in a number of fields, including 3D printing, there is no real knock-on effect in terms of employment, because services are not filling the gap. How, then, do we reconcile the different logics at work in national innovation – the business logic, the scientific and technological logics, the relationship to knowledge and the citizen logic of the democratic process – to achieve an innovation in which the different levels mutually enrich each other? These can be immersive innovations, which know how to ask the right questions, produce prototypes very quickly and multiply tests (which is precisely what additive manufacturing provides). But these best practices (immersion, questioning, creativity, prototyping and testing) are the cornerstones of design-thinking and are too little practiced in France (Ojasalo and Kauppinen 2016).

In today's competitive economy, we need to maintain "risky" investments that will eventually enable us to reduce the time scales between design and production on high-growth subjects such as those discussed in this book. The new deal may be to integrate aspects linked to the limits of reserves but, for the most part, we are not there yet! Whichever scenario emerges, there is no reason to systematically oppose support for the quest to satisfy disposable whims, the fear of others and the protection of someone's life and property, to the major degrowth of the "sober society". We need to take better account of the major trends in 3D printing research processes, if only to anticipate an inexorable future (provided we recognize the problems with appropriate means to limit them). For the future, we also need to rely on the ability of systems to evolve in order to maintain their functionalities in a world that will be increasingly disrupted (complexity, resilience, operation in degraded mode). But to do this, we need creative people!

Micnet (2020) suggests that, for each individual, there are several "temporal lines" all traced out. Thus, several "temporal paths" could be taken, following the laws of bifurcation managed in part by our brain and in part by external synchronicities. On this basis, we would ultimately retain a choice between these two types of interaction and our destiny would no longer be unilinear. As with bio-printing, it is a question of finding the right effectors for each individual to become creative!

> In other words, the more performance conditions prevail, the more complex participation in the research collective and, by the same token, to the reputation of the research unit. This reduction translates into forms of collective inaptitude affecting research staff, who are increasingly committed to individualism and to operating in disciplinary 'silos' (because they are evaluated by peers according to known methods). These various reduction processes can destroy the notion of the incommensurability of knowledge and, if we're not careful, lead to the implementation of a cold, but common, quantification of the entire research apparatus, reaching bases of unreliability because they are impartial (André 2017c).

How can we identify the people in the group who will take the initial risks that will lead to the development of new associations that will be difficult for competitors to imitate? How can we support these non-conformists who are factors of scientific innovation to commit the unit to the new and the excellent? How can we take into account (i.e. support) the originality, creativity, and the ability to found a school (at least for a while)?

Disciplinary specialization is recognized as a necessity for deciphering the increasingly specific facets of the systems and to establish relationships between highly specific variables in different fields. However, the increasingly acute compartmentalization of approaches is becoming problematic, as it leads to a disconnection from concrete issues and is mostly confined to increasingly circumscribed groups who are the only ones to hear a highly specialized language. This form of organization, whatever may be said, continues to produce high-quality scientific output. However, with the financial subordination of research, links between researchers are weak (as is the relationship with the hierarchy). It is even possible for researchers to be isolated, except at times when they are looking for partnerships to respond to a call for tenders. In such cases, interdisciplinarity may be the only option.

But do we really need a framework to invent? What we have been able to show is that between the idea and its POC and application is a tough road, requiring self-sacrifice, perseverance and strong values. On the academic side, knowledge of the possibility of being better connected to the economic world through support for valorization is part of the rhetoric, but little of the reality. On the corporate side, this negative situation is not apparent when management has decided to rely on creation as a means of development. Does this mean that openings to society are not favored from the university side? In fact, it is often at the suggestion of industrialists that links are forged with research teams, because they know they will find the equipment and conceptual support they lack. They understand that entropy growth is irreversible only within closed systems. We need to work on systematically and effectively opening up the academic world to representatives from other worlds.

When the idea is simple, using known knowledge in an original way, and a POC does not require funding (see 3D printing), there are opportunities to go further. In fact, this is what has been shown. On the other hand, for slightly more complicated (or complex) situations, weaknesses in interdependencies limit, as is the case with 4D printing and bio-printing, their promotion. Given the complexity processes associated with the idea, interdisciplinarity is a *sine qua non* for identifying and making sense of complex systems. It is a specific research approach built around one object, the idea. The modalities of this integration are neither clear nor simple to implement, and raise a host of questions that are addressed in this book.

Consequently, interdisciplinarity refers to institutional and intellectual issues (interdisciplinarity is conceived as the knowledge capable of grasping "what is woven together", that is to say, according to the original meaning of the term, the "complex" and goes against "knowledge which separates, breaks the complex of the world into disjointed fragments, and which splits the problems"), and epistemological (interdisciplinarity comes brutally to question science, at least the conception epistemology gives) (Morin 1998). Working in "interdisciplinarity" means building shared expertise. Interdisciplinarity is a work process and not an end in itself. In this sense, knowledge of the disciplines and of those who can represent them is essential to the proper functioning of interdisciplinarity, if one does not want to develop a superficial view of the phenomena studied. But, as has been shown, several forms of expertise are sometimes necessary with varying degrees of depth.

Finally, a mutual adjustment to each other's working methods seemed necessary, with points of friction and divergence appearing between the different approaches.

Generally speaking, innovators, inventors and creative people only come forward in situations where there is sufficient uncertainty to be able to exercise a scientific and/or technical skill which is not "regulatory" vested in them (this is possible in academic research). And to be the object of an active appropriation, these spaces of intellectual play need to be inhabited by players with a level of scientific and/or technical competence that is higher and/or different from that for which they were "placed" in the job. Moreover, apart from the essential original idea coming from motivated people, will we have the necessary funding (and especially people) to take the risk of pursuing a new approach which can, moreover, be interdisciplinary? Without this, scientific uncertainty, a normal situation in research, cannot be transformed into a space for creativity.

These aspects, which cannot ignore increasingly contractual financing methods, can limit creativity through unsuitable financial support frameworks and/or through carelessness or lack of support for risk taking. However, in the long term, it is of course on this upstream risk-taking, individual in its birth, then more collective in supervisory and scientific and technological support groups, that its relevance will be judged by the researcher or their research unit! This observation means that the organization of the laboratory must have reflective and united partners who anticipate change, who are not only part of "protection values" (satisfying passive precepts of biased evaluation, for example), and in the hierarchical logics of an old order and ordinary careers.

In this context, science (researchers and organizations) must acknowledge its social responsibility. It must therefore reflect on its aims and consequences. It must include a reflection on what technological progress represents, which is not an end in itself, but a means to (good) living on our planet. By reflecting on its aims, it must not claim to be neutral for the sole reason of asserting its objectivity, but be committed to better supporting the sharing of knowledge and the applicative powers linked to this knowledge.

Let us remember that "knowledge is the only matter that increases when shared" (Socrates, quoted by Mazurek (2016)). So, even if the decisive scientific discovery is an exception in the daily routine of research, remember to get out of your comfort zones! Be creative!

> Men are mistaken in thinking themselves free; their opinion is made up of consciousness of their own actions, and ignorance of the causes by which they are conditioned. Their idea of freedom, therefore is simply their ignorance of any cause for their actions (Spinoza 1994).

No administrative or scholastic nurturing will replace the miraculous opportune chances to which the world owes its great men (de Balzac 1973).

Being different is neither a good thing nor a bad thing; it simply means you're brave enough to be yourself (Gainnier 2020).

C.3. References

André, J.C. (2017a). *From Additive Manufacturing to 3D/4D Printing 1: From Concepts to Achievements*. ISTE Ltd, London, and John Wiley & Sons, New York.

André, J.C. (2017b). *From Additive Manufacturing to 3D/4D Printing 2: Current Techniques, Improvements and their Limitations*. ISTE Ltd, London, and John Wiley & Sons, New York.

André, J.C. (2017c). *From Additive Manufacturing to 3D/4D Printing 3: Breakthrough Innovations: Programmable Material, 4D Printing and Bio-printing*. ISTE Ltd, London, and John Wiley & Sons, New York.

de Balzac, H. (1973). *Cousin Pons*. Paperback, Paris.

Besnier, J.M. (2013). Seul le désordre est créateur. Pour en finir avec les bataillons disciplinaires. *La Revue Hermès*, 67, 25–31.

Colombo, M.G., von Krogh, G., Rossi-Lamastra, C., Stephan, P.E. (2017). Organizing for radical innovation: Exploring novel insights. *Journal of Product Innovation Management*, 34, 394–405.

Cowen, T. (2011). *The Great Stagnation: How American Are All the Low-Hanging Fruit of Modern History, Got Sick and Will (Eventually) Feel Better*. Dutton, New York.

Dupuy, J.P. (2002). *Pour un catastrophisme éclairé ; quand l'impossible est certain*. Le Seuil, Paris.

Encyclopedia of Chicago (1934). Century of Progress Exposition [Online]. Available at: http://www.encyclopedia.chicagohistory.org/pages/225.html.

Gainnier, B. (2020). *Chefs d'entreprise, ce que le monde attend de nous*. Alisio, Paris.

Hamel, J. (2002). La pédagogie comme pivot de l'interdisciplinarité. *Revue internationale d'éducation de Sèvres*, 30, 143–151.

Leroux, T. (2011). *Le laboratoire des pollutions industrielles : Paris 1770-1830*. Albin Michel, Paris.

Mazurek, H. (2016). Le partage des savoirs. Un nouveau paradigme en sciences sociales ? [Online]. Available at: https://horizon.documentation.ird.fr/exl-doc/pleins_textes/divers16-02/010066392.pdf.

Micnet, (2020). Libre-arbitre ou déterminisme : que dit la science ? [Online]. Available at: https://www.agoravox.tv/tribune-libre/article/libre-arbitre-ou-determinisme-que-85395.

Moreno, R. (1990). *Théorie du bordel ambiant. Souvenirs de l'irréversible*. Belfond, Paris.

Morin, E. (1998). Réformons la pensée. *Le Monde de l'éducation*, 252, 28.

Morin, E. (1999). Affronter l'incertitude. *Sciences humaines. Hors-série : la dynamique des savoirs*, 24, 66–67.

Naudon, F. (2013). Comment le profane joue en faveur du décloisonnement. *La Revue Hermès*, 67, 62–67.

Nicholas, J., Ledwith, A., Bessant, J. (2015). Selecting early-stage ideas for radical innovation: Tools and structures. *Research Technology Management*, 58, 36–44.

Ojasalo, J. and Kauppinen, H. (2016). Collaborative innovation with external actors: An empirical study on open innovation platforms in smart cities. *Technology Innovation Management Review*, 49–60 [Online]. Available at: https://timreview.ca/sites/default/files/article_PDF/OjasaloKauppinen_TIMReview_December2016.pdf.

Serres, M. (1969). *La communication*. Les Éditions de Minuit, Paris.

Sharp, G. (1995). *La guerre civilisée. La défense par actions civiles*. Presses universitaires de Grenoble, Grenoble.

Sonigo, P. (2000). Cellules en liberté. In *Ni Dieu, ni gène ; pour une autre théorie de l'hérédité*, Kupiec, J.J. and Sonigo, P. (eds). Points Sciences, Paris.

Spinoza, B. (1994). *The Ethics*. Folio Essais, Paris.

Thill, G. (1987). Interdisciplinarité et enseignement supérieur, Point de vue européen. *UNESCO – Cahiers Bibliosur l'enseignement supérieur*, 24, 19–35.

Trégouët, R. (2020). Le développement durable et l'innovation globale pourraient nous permettre d'affronter les défis de ce siècle [Online]. Available at: https://www.rtflash.fr/newsletter.

Index

A, B

acceptability, 3, 4, 9, 12, 31, 32, 38, 49, 54, 61, 141
acceptance, 124
accidents, 16, 22, 31–34, 37
actuation, 129
additive manufacturing, 92–98, 100–104, 107, 109–112, 118, 120, 132, 135, 139, 143, 186–191, 198, 200, 203, 216, 221
ALARA, 43, 47
alert, 43
anarchism, 165
anticipation, 29, 174
anxiety, 20, 35
application, 14, 42, 47, 50, 56, 59, 62, 91, 94, 115, 145, 199, 200, 202, 221, 223
artifact, 23
attractiveness, 12, 105, 128, 189
attractors, 136, 137
autonomy, 24, 37, 39, 50, 64, 67, 93, 181
ban, 49
belief, 11, 106
bias, 35, 36, 47, 48, 92, 104, 174, 181
bifurcation, 42, 64, 136–138, 222
Big Bang, 113, 162
bio-printing, 93, 94, 117, 129–135, 138–143, 146, 163, 186, 187, 198, 199, 202, 204, 222, 223
bisociation, 145
bovine spongiform encephalopathy, 55

C, D

causal thinking, 199
cause-effect, 21
cells, 130–135, 138, 139, 143, 198–200, 215
change(s), 19, 24, 33, 66, 70, 92, 96, 97, 121, 128, 136, 139, 167, 171, 172, 175, 177, 191, 194, 224
clarification, 50, 51, 99, 139, 168, 170–174, 177–181, 184, 185, 192
collective, 46, 61, 64, 103, 117, 177, 222, 224
commitment, 6, 64, 65, 96, 106, 124, 127, 141, 175, 182, 185
compatibilities, 133
competence, 10, 21, 52, 60, 125, 126, 224

complexity, 15, 19, 22, 27–29, 32, 33, 36, 50, 59, 69, 70, 110, 125, 132, 135, 139–141, 144, 174–177, 181–184, 191, 197, 200, 202, 203, 222, 223
conceptual hurdle, 204
confused apprehension, 162, 163, 167, 171, 174, 179
conservatism, 118, 220
consistency, 49, 54, 177, 179
constraints, 2, 5, 12, 13, 18, 20, 36, 51, 60, 67, 109, 110, 114, 122, 132, 135, 170, 172, 175, 180, 183, 185, 190, 191, 216
control, 9, 15, 22, 32, 33, 35–38, 40, 41, 45, 49, 53, 68, 123, 136, 219
convergence, 38, 95, 102, 119, 140, 202
creative destruction, 67, 93, 95
creativity, 14, 18, 72, 92, 94, 95, 107, 111, 116, 125, 127, 143–145, 162–165, 168, 169, 185, 187, 188, 203, 215–217, 221, 222, 224
crises, 29–32, 35, 36, 38, 62, 66
culture, 5–7, 14, 15, 17, 25, 26, 33, 39, 41, 48, 50, 60, 61, 63, 65, 66, 96, 106, 143, 144, 162, 168, 169, 177, 199, 203
dangers/hazards, 30, 33, 47, 49, 54, 61
dead ends/impasses, 171, 177
death valley, 67
decision-making, 9–12, 16, 20, 23, 37, 42, 52–54, 58, 60, 62, 63, 65, 70, 105, 106, 112, 124, 174, 175, 177, 180, 181
deepening, 16, 42, 51, 59, 143, 174
democracy, 4, 23, 40, 51, 59, 62, 67
deontology, 13
design, 6, 8, 11, 12, 24, 37, 59, 61, 67, 96, 98–102, 106, 110, 111, 122, 124, 133, 184, 187, 190, 221
 -thinking, 128, 221
determinism, 118, 145, 199, 200
difficulties, 22, 50, 59, 99, 101, 104, 105, 111, 122, 130, 143, 163, 170, 173, 180, 181, 185, 202
digital, 2, 47, 100, 124
Direct Energy Deposition (DED), 98
disciplinary hybridization, 104, 140
disorder, 20, 68, 142, 144
divergence, 69, 95, 115, 145, 178, 183, 224
DIY, 116
dream, 17, 102, 143, 146
dynamic, 8, 10, 24, 28, 37, 38, 48, 63, 93, 104, 107, 111, 126, 145, 146, 164, 181, 204

E, F

economy, 5, 32, 65, 67, 91, 101, 104, 124, 126, 182, 203, 215, 221
emergence, 3, 31, 32, 66, 91, 93, 104, 125, 144, 163, 167, 185, 189, 204, 221
emissions, 2, 51, 59
energetic stimuli, 122
epistemology, 17, 106, 140, 202, 223
ethics, 2, 3, 5, 6, 8, 11–13, 16, 17, 33, 40, 43, 46–48, 51, 59, 63, 70, 106, 133, 164, 181, 215
Eureka!, 168, 170
evaluation, 14, 28, 48, 58–60, 64, 95, 98, 115, 124, 141, 170, 174, 180, 184, 185, 216, 224
expertise/expert, 31–33, 46, 50, 56–59, 62, 67, 171, 172, 174, 176–181, 184
fatigue, 122
forward-looking, 28, 46, 69
fragility, 13, 30, 181
framework, 4, 6, 7, 13, 14, 18–20, 25, 28, 29, 34, 42, 49–52, 59, 60, 62, 63, 68, 69, 71, 96, 114, 164, 165, 172–174, 179, 180, 219, 223, 224

freedom, 3, 8, 12, 15, 17, 20, 29, 61, 64, 68, 93, 100, 101, 111, 116, 123, 124, 140, 143, 144, 162, 168, 190, 220, 224

G, H, I

global approach, 135, 219
hacker, 127
HCSP, 55–57
heterogeneity, 95, 133, 146
heuristic, 52, 69, 103, 124, 174, 184, 200
illnesses/diseases, 22, 31–34
imagination, 17, 68, 92, 95, 125, 129, 144, 161, 164
improvement, 21, 42, 46, 96, 105, 142, 165
indecision, 42
independence, 10, 38, 165
information, 23, 32–36, 47, 49, 51, 53, 63, 101, 117, 127, 163, 173, 187, 190
initial conditions, 139
innovation, 2–4, 6, 9–11, 16, 19, 28, 37, 40–42, 44, 51, 59, 63, 67–72, 91, 95, 99, 102–104, 107, 109, 111, 113, 114, 116, 117, 124, 125, 127–129, 140, 143, 145, 146, 162, 165, 175, 181, 182, 185, 187–189, 203, 204, 215–218, 221, 222
insecurity, 38, 183
institutes, 21, 29, 40
insurance, 31
integration, 9, 26, 40, 56, 69, 70, 103, 104, 114, 115, 131, 175, 176, 179, 223
integrity, 14, 51, 52
interdependencies, 13, 65, 135, 139, 197, 223
interdisciplinarity, 17, 19, 26, 29, 68, 105, 126, 129, 140, 162, 163, 165, 173, 176–179, 183, 184, 202, 218, 219, 223
interest(s), 3, 10, 12, 24, 48, 51, 59, 65, 99, 102, 106, 127, 174
invention, 15, 18, 24, 40, 66, 67, 69, 70, 91–93, 95, 96, 99, 104, 107, 109, 115, 116, 128, 133, 140, 145, 161–164, 171, 174, 180, 184–189, 203, 215–218, 221
irrationality, 64, 167
irreversibility, 8, 18, 21, 43, 49, 54

K, L, M

knowledge, 3, 6, 10, 12, 16–19, 21, 22, 28, 30, 32–34, 36, 37, 49–53, 59, 62, 64, 95, 96, 106, 117, 123, 126, 130, 144, 163, 164, 166, 175, 177, 178, 190, 192, 197, 199, 218, 223
law, 53
learning, 6, 48, 104, 141, 167, 171, 182, 193, 196, 202
legal, 14, 37, 45, 63
legitimacy, 4, 9, 16, 25, 37, 95, 102, 146, 180, 182
life/living, 6, 51, 59, 63, 93, 117, 132–140, 142, 187, 189, 198–200, 202, 204
limit values, 24, 56
limitations, 129, 133
limits, 21, 22, 35, 52, 60, 96, 101, 102, 122, 126, 127, 132, 143, 166, 179, 184, 190, 200, 220, 221
linear process, 28, 111, 121
malaise, 4, 18
mapping, 49, 54, 55, 57, 58
marginalization, 35, 185
material extrusion, 98
media coverage, 40

methodology, 29, 120
mimicry, 220
mobile telephony, 55
mobilization/mobilizing, 3, 24, 52, 183
modeling, 14, 104, 119, 120, 122, 181, 199
moratorium, 49, 53
morphogenesis, 126, 191, 200

N, O, P

nanoparticles, 31, 55–57, 61, 120
niches, 117, 122, 139, 196, 197, 200, 202
nonlinear dynamical systems (NLDS), 136
objects, 5, 15, 61, 93, 96, 97, 99–101, 104, 105, 117, 118, 122, 123, 132, 135, 141, 183, 187, 188, 191, 192, 196, 198, 217, 221
observability, 49, 54, 70
ontology, 176
opinion, 10, 23, 25, 27, 40, 42, 61, 224
opportunity, 183, 185, 202
organ, 132, 133, 198
pandemic, 29, 38, 58
participation, 9, 23, 109, 188, 222
Pasteur's quadrant, 114, 115
performance, 20, 25, 26, 43, 45, 70, 182, 203, 217, 218, 222
philosophy, 20, 69, 94, 106, 167
pioneering spirit, 112
plagiarism, 14, 51, 71
plausibility, 49, 54
POC, 6, 61, 69, 70, 91, 92, 99, 103, 123
political/policy, 4, 12, 23, 25, 29, 43, 51, 59, 65, 104, 107, 128, 144, 182, 184
pollution, 51, 59, 194
pooling of knowledge, 142, 177
powder bed fusion, 98
power of the unpredictable, 140
precariousness, 19, 31
pre-POC, 43, 45, 46, 61, 63, 64
prevention, 21, 23, 26, 30, 34, 36, 37, 51, 57, 59
principals, 28, 40, 63
principle, 3, 5, 8, 22, 25, 26, 28, 33, 35–37, 40, 41, 44, 45–47, 49, 53, 55, 61, 62, 93, 97, 109, 118, 129–134, 136, 141, 143, 144, 162, 175–177, 187, 188, 191, 198, 200, 204
printability, 188, 189
processes, 24, 37, 61, 70, 92, 98, 100, 101, 105, 107–110, 115, 117, 119, 120, 123, 135, 185, 187–189, 191, 192, 200, 222
professionalism, 165
profusion, 144
progress, 3, 6, 8, 16–20, 26, 32, 36, 42, 44, 47, 51, 59, 61, 62, 71, 101, 102, 125, 128, 180, 203, 215–217, 220, 224
project, 42, 69, 96, 106, 112, 124, 127, 138, 144, 172, 180, 183, 202, 217

R, S

REACH, 56, 57, 61
recommendations, 56, 57, 60
reducibility, 49, 54
regulations, 22, 26, 34, 43, 45, 53, 70, 127
repellents, 137
representation, 12, 22, 25, 30, 58, 117, 124, 137, 166, 167, 171, 176, 184
reproducibility, 14, 51, 52, 122, 137
respect, 14, 48, 109, 185
responsibility, 2, 3, 6, 9, 12, 14, 17, 28, 30, 35–38, 40, 41, 43–48, 62–65, 67, 141, 144, 146, 175, 215, 224
restriction, 49, 53
revolution, 20, 71, 101, 103, 187, 204, 219, 220
rhetorical, 22, 66

risk management, 29, 47, 62
robotics, 117, 129, 193, 196
robustness, 14, 117, 122, 173, 181, 197
scaffolding, 133, 134, 139
science, 1, 3, 4, 9, 14–18, 22–26, 28, 29, 39, 43, 45, 47, 48, 53, 62, 65, 93, 94, 96, 98, 102, 106, 107, 114, 116, 126, 133, 136, 141, 142, 165, 166, 178, 180, 197, 200, 216–220, 223, 224
scientific
  approach, 92
  rationality, 16, 202
severity, 49, 54
sheet lamination, 98
silo, 40, 65, 222
singularities, 103, 119
social demand, 22, 24, 33, 40
socialization, 12, 20, 66, 140
socially responsible research (SSR), 42
society, 2–9, 11, 12, 15–33, 36, 38–40, 42–46, 49, 50, 53, 58–62, 64–66, 68, 70, 72, 94, 104, 106, 107, 116, 128, 130, 141, 144–146, 166, 197, 202, 220–223
spatial distributions, 135, 136, 200
specialization, 93, 165, 178, 222
stakeholders, 4, 21, 30, 40, 44, 47, 48, 50, 58, 59, 61, 62, 66, 126, 127
standards, 44
subsidiarity, 70, 104
suspicion, 32, 35, 36
synthesis, 120, 123, 139, 141, 171, 178
systems, 6, 11, 12, 21, 29, 30, 32, 36, 40, 50, 53, 58, 59, 69, 93, 101, 104, 110–114, 117, 126, 128, 131, 135, 136, 138–140, 144, 165, 175, 178, 181, 191, 195, 197, 198, 200, 217, 222, 223

T, U

technological paradigm, 126
technology, 4, 5, 12, 15, 17–19, 27, 29, 40, 41, 48, 62, 65, 69, 70, 93, 97, 101, 102, 109–111, 115–118, 120, 122, 126–128, 132, 142, 146, 166, 186–188, 197, 200, 204
Technology Readiness Levels (TRL), 6, 7
time pressure, 25, 62, 116
tinkering, 123, 126, 168, 174, 219
tipping over, 167
tissue, 93, 132–134, 198
traceability, 43, 123, 185
training, 14, 17, 20, 60, 69, 114, 133, 164, 177, 185
transplants, 132
trivialization, 29
trust/confidence, 4, 9, 14, 18, 21, 24–26, 30, 32, 36–38, 43, 46, 64, 66, 104, 111, 112, 129, 146, 175, 180, 187, 217
uncertain, 22, 45, 56, 144, 197
uncertainty, 12, 22, 23, 31, 37, 49, 53, 54, 58, 68, 116, 124, 184, 224
unemployment, 31
unknown futures, 19
utopia, 20, 177

V, W

victims, 30, 33, 35–37
violence, 30–32
vitalism, 200
well-being, 2, 12, 61, 67

Other titles from

ISTE

in

Systems and Industrial Engineering – Robotics

2023

BERRAH Lamia, CLIVILLÉ Vincent
Decision and Decision-maker in an Industrial Environment

BRIFFAUT Jean-Pierre
Complexities 1: Various Approaches in the Field of Techno-scientific Knowledge
(Systems of Systems Complexity – Volume 5)

CHAINEAUX Jacques
The Mechanisms of Explosions: 27 Case Studies for their Understanding

DIALLO Saikou, BOGUI Jean-Jacques Maomra, ATCHOUA Julien
Technologies and African Societies in Pandemic Times: Using Technology to Survive and Thrive During the Covid-19 Era
(Africa Digital Age Set – Volume 2)

JAEGLER Anicia, ROQUES Thierry
Sustainable Supply Chain Management: Sustainable Decision Support

PRINTZ Jacques
Organization and Pedagogy of Complexity: Systemic Case Studies and Prospects
(Systems of Systems Complexity Set – Volume 4)

2022

AMARA Yacine, BEN AHMED Hamid, GABSI Mohamed
Hybrid Excited Synchronous Machines: Topologies, Design and Analysis

BOURRIÈRES Jean-Paul, PINÈDE Nathalie, TRAORÉ Mamadou Kaba, ZACHAREWICZ Grégory
From Logistic Networks to Social Networks: Similarities, Specificities, Modeling, Evaluation

DEMOLY Frédéric, ANDRÉ Jean-Claude
4D Printing 1: Between Disruptive Research and Industrial Applications
4D Printing 2: Between Science and Technology

HAJJI Rafika, JARAR OULIDI Hassane
Building Information Modeling for a Smart and Sustainable Urban Space

KROB Daniel
Model-based Systems Architecting: Using CESAM to Architect Complex Systems
(Systems of Systems Complexity Set – Volume 3)

LOUIS Gilles
Dynamics of Aircraft Flight

2020

BRON Jean-Yves
System Requirements Engineering

KRYSINSKI TOMASZ, MALBURET FRANÇOIS
Energy and Motorization in the Automotive and Aeronautics Industries

PRINTZ Jacques
System Architecture and Complexity: Contribution of Systems of Systems to Systems Thinking

2019

ANDRÉ Jean-Claude
Industry 4.0: Paradoxes and Conflicts

BENSALAH Mounir, ELOUADI Abdelmajid, MHARZI Hassan
Railway Information Modeling RIM: The Track to Rail Modernization

BLUA Philippe, YALAOU Farouk, AMODEO Lionel, DE BLOCK Michaël, LAPLANCHE David
Hospital Logistics and e-Management: Digital Transition and Revolution

BRIFFAUT Jean-Pierre
From Complexity in the Natural Sciences to Complexity in Operations Management Systems
(Systems of Systems Complexity Set – Volume 1)

BUDINGER Marc, HAZYUK Ion, COÏC Clément
Multi-Physics Modeling of Technological Systems

FLAUS Jean-Marie
Cybersecurity of Industrial Systems

JAULIN Luc
Mobile Robotics – Second Edition Revised and Updated

KUMAR Kaushik, DAVIM Paulo J.
Optimization for Engineering Problems

TRIGEASSOU Jean-Claude, MAAMRI Nezha
Analysis, Modeling and Stability of Fractional Order Differential Systems 1: The Infinite State Approach
Analysis, Modeling and Stability of Fractional Order Differential Systems 2: The Infinite State Approach

VANDERHAEGEN Frédéric, MAAOUI Choubeila, SALLAK Mohamed, BERDJAG Denis
Automation Challenges of Socio-technical Systems

2018

BERRAH Lamia, CLIVILLÉ Vincent, FOULLOY Laurent
Industrial Objectives and Industrial Performance: Concepts and Fuzzy Handling

GONZALEZ-FELIU Jesus
Sustainable Urban Logistics: Planning and Evaluation

GROUS Ammar
Applied Mechanical Design

LEROY Alain
Production Availability and Reliability: Use in the Oil and Gas Industry

MARÉ Jean-Charles
Aerospace Actuators 3: European Commercial Aircraft and Tiltrotor Aircraft

MAXA Jean-Aimé, BEN MAHMOUD Mohamed Slim, LARRIEU Nicolas
Model-driven Development for Embedded Software: Application to Communications for Drone Swarm

MBIHI Jean
Analog Automation and Digital Feedback Control Techniques
Advanced Techniques and Technology of Computer-Aided Feedback Control

MORANA Joëlle
Logistics

SIMON Christophe, WEBER Philippe, SALLAK Mohamed
Data Uncertainty and Important Measures
(Systems Dependability Assessment Set – Volume 3)

TANIGUCHI Eiichi, THOMPSON Russell G.
City Logistics 1: New Opportunities and Challenges
City Logistics 2: Modeling and Planning Initiatives
City Logistics 3: Towards Sustainable and Liveable Cities

ZELM Martin, JAEKEL Frank-Walter, DOUMEINGTS Guy, WOLLSCHLAEGER Martin
Enterprise Interoperability: Smart Services and Business Impact of Enterprise Interoperability

2017

ANDRÉ Jean-Claude
From Additive Manufacturing to 3D/4D Printing 1: From Concepts to Achievements
From Additive Manufacturing to 3D/4D Printing 2: Current Techniques, Improvements and their Limitations
From Additive Manufacturing to 3D/4D Printing 3: Breakthrough Innovations: Programmable Material, 4D Printing and Bio-printing

ARCHIMÈDE Bernard, VALLESPIR Bruno
Enterprise Interoperability: INTEROP-PGSO Vision

CAMMAN Christelle, FIORE Claude, LIVOLSI Laurent, QUERRO Pascal
Supply Chain Management and Business Performance: The VASC Model

FEYEL Philippe
Robust Control, Optimization with Metaheuristics

MARÉ Jean-Charles
Aerospace Actuators 2: Signal-by-Wire and Power-by-Wire

POPESCU Dumitru, AMIRA Gharbi, STEFANOIU Dan, BORNE Pierre
Process Control Design for Industrial Applications

RÉVEILLAC Jean-Michel
Modeling and Simulation of Logistics Flows 1: Theory and Fundamentals
Modeling and Simulation of Logistics Flows 2: Dashboards, Traffic Planning and Management
Modeling and Simulation of Logistics Flows 3: Discrete and Continuous Flows in 2D/3D

2016

ANDRÉ Michel, SAMARAS Zissis
Energy and Environment
(Research for Innovative Transports Set – Volume 1)

AUBRY Jean-François, BRINZEI Nicolae, MAZOUNI Mohammed-Habib
Systems Dependability Assessment: Benefits of Petri Net Models
(Systems Dependability Assessment Set – Volume 1)

BLANQUART Corinne, CLAUSEN Uwe, JACOB Bernard
Towards Innovative Freight and Logistics
(Research for Innovative Transports Set – Volume 2)

COHEN Simon, YANNIS George
Traffic Management
(Research for Innovative Transports Set – Volume 3)

MARÉ Jean-Charles
Aerospace Actuators 1: Needs, Reliability and Hydraulic Power Solutions

REZG Nidhal, HAJEJ Zied, BOSCHIAN-CAMPANER Valerio
Production and Maintenance Optimization Problems: Logistic Constraints and Leasing Warranty Services

TORRENTI Jean-Michel, LA TORRE Francesca
Materials and Infrastructures 1
(Research for Innovative Transports Set – Volume 5A)
Materials and Infrastructures 2
(Research for Innovative Transports Set – Volume 5B)

WEBER Philippe, SIMON Christophe
Benefits of Bayesian Network Models
(Systems Dependability Assessment Set – Volume 2)

YANNIS George, COHEN Simon
Traffic Safety
(Research for Innovative Transports Set – Volume 4)

2015

AUBRY Jean-François, BRINZEI Nicolae
Systems Dependability Assessment: Modeling with Graphs and Finite State Automata

BOULANGER Jean-Louis
CENELEC 50128 and IEC 62279 Standards

BRIFFAUT Jean-Pierre
E-Enabled Operations Management

MISSIKOFF Michele, CANDUCCI Massimo, MAIDEN Neil
Enterprise Innovation

2014

CHETTO Maryline
Real-time Systems Scheduling
Volume 1 – Fundamentals
Volume 2 – Focuses

DAVIM J. Paulo
Machinability of Advanced Materials

ESTAMPE Dominique
Supply Chain Performance and Evaluation Models

FAVRE Bernard
Introduction to Sustainable Transports

GAUTHIER Michaël, ANDREFF Nicolas, DOMBRE Etienne
Intracorporeal Robotics: From Milliscale to Nanoscale

MICOUIN Patrice
Model Based Systems Engineering: Fundamentals and Methods

MILLOT Patrick
Designing Human–Machine Cooperation Systems

NI Zhenjiang, PACORET Céline, BENOSMAN Ryad, RÉGNIER Stéphane
Haptic Feedback Teleoperation of Optical Tweezers

OUSTALOUP Alain
Diversity and Non-integer Differentiation for System Dynamics

REZG Nidhal, DELLAGI Sofien, KHATAD Abdelhakim
Joint Optimization of Maintenance and Production Policies

STEFANOIU Dan, BORNE Pierre, POPESCU Dumitru, FILIP Florin Gh.,
EL KAMEL Abdelkader
Optimization in Engineering Sciences: Metaheuristics, Stochastic Methods and Decision Support

2013

ALAZARD Daniel
Reverse Engineering in Control Design

ARIOUI Hichem, NEHAOUA Lamri
Driving Simulation

CHADLI Mohammed, COPPIER Hervé
Command-control for Real-time Systems

DAAFOUZ Jamal, TARBOURIECH Sophie, SIGALOTTI Mario
Hybrid Systems with Constraints

FEYEL Philippe
Loop-shaping Robust Control

FLAUS Jean-Marie
Risk Analysis: Socio-technical and Industrial Systems

FRIBOURG Laurent, SOULAT Romain
Control of Switching Systems by Invariance Analysis: Application to Power Electronics

GROSSARD Mathieu, RÉGNIER Stéphane, CHAILLET Nicolas
Flexible Robotics: Applications to Multiscale Manipulations

GRUNN Emmanuel, PHAM Anh Tuan
Modeling of Complex Systems: Application to Aeronautical Dynamics

HABIB Maki K., DAVIM J. Paulo
Interdisciplinary Mechatronics: Engineering Science and Research Development

HAMMADI Slim, KSOURI Mekki
Multimodal Transport Systems

JARBOUI Bassem, SIARRY Patrick, TEGHEM Jacques
Metaheuristics for Production Scheduling

KIRILLOV Oleg N., PELINOVSKY Dmitry E.
Nonlinear Physical Systems

LE Vu Tuan Hieu, STOICA Cristina, ALAMO Teodoro,
CAMACHO Eduardo F., DUMUR Didier
Zonotopes: From Guaranteed State-estimation to Control

MACHADO Carolina, DAVIM J. Paulo
Management and Engineering Innovation

MORANA Joëlle
Sustainable Supply Chain Management

SANDOU Guillaume
Metaheuristic Optimization for the Design of Automatic Control Laws

STOICAN Florin, OLARU Sorin
Set-theoretic Fault Detection in Multisensor Systems

2012

AÏT-KADI Daoud, CHOUINARD Marc, MARCOTTE Suzanne, RIOPEL Diane
Sustainable Reverse Logistics Network: Engineering and Management

BORNE Pierre, POPESCU Dumitru, FILIP Florin G., STEFANOIU Dan
Optimization in Engineering Sciences: Exact Methods

CHADLI Mohammed, BORNE Pierre
Multiple Models Approach in Automation: Takagi-Sugeno Fuzzy Systems

DAVIM J. Paulo
Lasers in Manufacturing

DECLERCK Philippe
Discrete Event Systems in Dioid Algebra and Conventional Algebra

DOUMIATI Moustapha, CHARARA Ali, VICTORINO Alessandro, LECHNER Daniel
Vehicle Dynamics Estimation using Kalman Filtering: Experimental Validation

GUERRERO José A, LOZANO Rogelio
Flight Formation Control

HAMMADI Slim, KSOURI Mekki
Advanced Mobility and Transport Engineering

MAILLARD Pierre
Competitive Quality Strategies

MATTA Nada, VANDENBOOMGAERDE Yves, ARLAT Jean
Supervision and Safety of Complex Systems

POLER Raúl, CARNEIRO Luis Maia, JASINSKI Thomas, ZOLGHADRI Marc, PEDRAZZOLI Paolo
Intelligent Non-hierarchical Manufacturing Networks

TROCCAZ Jocelyne
Medical Robotics

YALAOUI Alice, CHEHADE Hicham, YALAOUI Farouk, AMODEO Lionel
Optimization of Logistics

ZELM Martin, SANCHIS Raquel, POLER Raúl, DOUMEINGTS Guy
Enterprise Interoperability –I-EASA12 Proceedings

2011

CANTOT Pascal, LUZEAUX Dominique
Simulation and Modeling of Systems of Systems

DAVIM J. Paulo
Mechatronics

DAVIM J. Paulo
Wood Machining

GROUS Ammar
Applied Metrology for Manufacturing Engineering

KOLSKI Christophe
Human–Computer Interactions in Transport

LUZEAUX Dominique, RUAULT Jean-René, WIPPLER Jean-Luc
Complex Systems and Systems of Systems Engineering

ZELM Martin, VAN SINDEREN Marten, DOUMEINGTS Guy, JOHNSON Pontus
Enterprise Interoperability: IWEI2011 Proceedings

2010

BOTTA-GENOULAZ Valérie, CAMPAGNE Jean-Pierre, LLERENA Daniel, PELLEGRIN Claude
Supply Chain Performance: Collaboration, Alignment and Coordination

BOURLÈS Henri, GODFREY K.C. Kwan
Linear Systems

BOURRIÈRES Jean-Paul
Proceedings of CEISIE'09

CHAILLET Nicolas, REGNIER Stéphane
Microrobotics for Micromanipulation

DAVIM J. Paulo
Sustainable Manufacturing

GIORDANO Max, MATHIEU Luc, VILLENEUVE François
Product Life-Cycle Management: Geometric Variations

LOZANO Rogelio
Unmanned Aerial Vehicles: Embedded Control

LUZEAUX Dominique, RUAULT Jean-René
Systems of Systems

VILLENEUVE François, MATHIEU Luc
Geometric Tolerancing of Products

2009

DIAZ Michel
Petri Nets: Fundamental Models, Verification and Applications

OZEL Tugrul, DAVIM J. Paulo
Intelligent Machining

PITRAT Jacques
Artificial Beings

2008

ARTIGUES Christian, DEMASSEY Sophie, NÉRON Emmanuel
Resources–Constrained Project Scheduling

BILLAUT Jean-Charles, MOUKRIM Aziz, SANLAVILLE Eric
Flexibility and Robustness in Scheduling

DOCHAIN Denis
Bioprocess Control

LOPEZ Pierre, ROUBELLAT François
Production Scheduling

THIERRY Caroline, THOMAS André, BEL Gérard
Supply Chain Simulation and Management

2007

DE LARMINAT Philippe
Analysis and Control of Linear Systems

DOMBRE Etienne, KHALIL Wisama
Robot Manipulators

LAMNABHI-LAGARRIGUE Françoise, LAGHROUCHE Salah, LORIA Antonio, PANTELEY Elena
Taming Heterogeneity and Complexity of Embedded Control

LIMNIOS Nikolaos
Fault Trees

2006

FRENCH COLLEGE OF METROLOGY
Metrology in Industry

NAJIM Kaddour
Control of Continuous Linear Systems

Printed and bound by CPI Group (UK) Ltd, Croydon, CR0 4YY
16/09/2026
14963082-0004